Wireless Networks

Series Editor

Xuemin Shen
University of Waterloo
Waterloo, ON, Canada

The purpose of Springer's new Wireless Networks book series is to establish the state of the art and set the course for future research and development in wireless communication networks. The scope of this series includes not only all aspects of wireless networks (including cellular networks, WiFi, sensor networks, and vehicular networks), but related areas such as cloud computing and big data. The series serves as a central source of references for wireless networks research and development. It aims to publish thorough and cohesive overviews on specific topics in wireless networks, as well as works that are larger in scope than survey articles and that contain more detailed background information. The series also provides coverage of advanced and timely topics worthy of monographs, contributed volumes, textbooks and handbooks.

More information about this series at http://www.springer.com/series/14180

Feng Lyu • Minglu Li • Xuemin Shen

Vehicular Networking for Road Safety

Springer

Feng Lyu
School of Computer Science
and Engineering
Central South University
Changsha, Hunan, China

Minglu Li
Computer Science and Engineering
Shanghai Jiao Tong University
Shanghai, China

Xuemin Shen ⓘ
Electrical and Computer Engineering
University of Waterloo
Waterloo, ON, Canada

ISSN 2366-1186   ISSN 2366-1445   (electronic)
Wireless Networks
ISBN 978-3-030-51231-6   ISBN 978-3-030-51229-3   (eBook)
https://doi.org/10.1007/978-3-030-51229-3

This Springer imprint is published by the registered company Springer Nature Switzerland AG.
The registered company address is: Gewerbestrasse 11, 6330 Cham, Switzerland

Preface

Road safety has always been the first priority for daily commuters. According to the World Health Organization (WHO), more than 1.2 million and 50 million people worldwide are killed and injured due to collisions each year, respectively. Being unresponsive to on-road emergencies is the major reason to most accidents, which poses the necessity of building active cooperative road-safety applications via Vehicular Ad hoc NETworks (VANETs). Particularly, empowered by vehicle-to-everything (V2X) communications, which broadly include vehicle-to-vehicle (V2V), vehicle-to-infrastructure (V2I), vehicle-to-pedestrian (V2P), etc., real-time environment information can be exchanged among neighboring vehicles rapidly via broadcasting road-safety beacons, i.e., cooperative awareness messages (CAMs), that can facilitate various advanced road-safety applications with pre-sensing capabilities. Compared with other sensors, such as camera, radar, and light detection and ranging (LiDAR), V2X communications can provide 360° situational awareness on road with offering more excellent sensing range, through-objects view functionality, and around-corner viewing capability. Besides, V2X communications are not affected or influenced by non-ideal weather conditions, such as heavy rain, fog, and harsh sunbeams, which can work robustly in real driving environments. Both advantages can collectively facilitate the extensive usage of V2X in future transportation systems, especially for road safety enhancement.

To well support road-safety applications, low-latency and reliable V2X communications are required, which however are challenging to be guaranteed under vehicular environments with fast-changing network topologies, intermittent wireless links, and dynamic traffic densities. First, at the MAC layer, as CAMs are related to road safety, minimizing the medium access delay and avoiding the medium access collision should be achieved simultaneously. However, due to the lack of a global central unit in vehicular environments, vehicles have to negotiate the medium access in a fully distributed way. Additionally, the fast-changing network topology can further render the MAC design intricate. Second, at the link layer, even granted with the appropriate medium resources, communication reliability remains to be further enhanced, as there are many uncontrollable factors, such as types of roads, time-varying traffic conditions, and all different surrounding buildings and trees, that can precariously affect the wireless link performance. Third, at the network layer, due to dynamic traffic densities, the naive broadcasting scheme with a fixed data rate and transmission power may cause severe channel congestion, especially under dense-vehicle scenarios, which can significantly degrade the V2X reliability.

In this monograph, we investigate vehicular networking technologies to guarantee low-latency and reliable V2X communications for road-safety applications. Specifically, we focus on dedicated short range communication (DSRC) technologies at the MAC, link, and network layers. In Chap. 1, we introduce vehicular networks, including its definition, technical challenges, and how it can support road-safety applications, etc. In Chap. 2, we review the state-of-the-art vehicular networking techniques and organize a comprehensive survey to state our technical motivations per the MAC, link, and network layer. In Chap. 3, to avoid medium access collisions caused by vehicular mobilities, we propose a mobility-aware TDMA-based MAC, named *MoMAC*, which can assign each vehicle a collision-avoidance time slot according to the underlying road topology and lane distribution on roads. As the existing vehicular MACs do not consider the situation that vehicles have diverse beaconing rates to support various road-safety applications, and such inflexible design may suffer from a scalability issue in terms of channel resource management, in Chap. 4, we propose a novel time slot-sharing MAC, named *SS-MAC*, to support diverse beaconing rates of vehicles. The proposed *MoMAC* and *SS-MAC* can work collectively to provide collision-free/reliable, scalable, and efficient medium access for moving and distributed vehicles. In Chap. 5, to understand the DSRC performance in urban environments, we implement a V2V communication testbed based on commodity onboard units (OBUs) and collect large volumes of beaconing traces together with the simultaneous environmental context information in Shanghai city, based on which we then conduct extensive data analytics to characterize the V2V communications. In Chap. 6, with the deep understanding on link characteristics, we propose a link-aware beaconing scheme, named *CoBe*, to enhance the broadcasting reliability by coping with harsh non-line-of-sight (NLoS) conditions. In Chap. 7, to adapt to dynamic vehicle densities with satisfying individual road safety demand, we further propose a fully distributed adaptive beaconing control scheme, named *ABC*, to conduct safety-aware beaconing rate adaptation for vehicles. At last, we conclude this monograph and provide potential future research issues in Chap. 8. The systematic principle in this monograph provides valuable guidance on the deployment and implementation of future VANET-enabled road-safety applications.

We would like to thank Prof. Hongzi Zhu at Shanghai Jiao Tong University, Dr. Haibo Zhou, Dr. Nan Cheng, Dr. Wenchao Xu, Dr. Huaqing Wu, and Dr. Haixia Peng from Broadband Communications Research (BBCR) Group at the University of Waterloo, for their contributions in the presented research works. We also would like to thank all the members of BBCR group for the valuable discussions and their insightful suggestions, ideas, and comments. In addition, I would personally thank my wife Ms. Xingxin Chen for her heartfelt support on my overseas studying and working and thank for the birth of my lovely son Haoran Lyu that directs and strengthens my faith in facing challenges. Special thanks also go to the staff at Springer Science+Business Media: Susan Lagerstrom-Fife, Shina Harshavardhan, and Christiane Bauer for their help throughout the publication preparation process.

Changsha, Hunan, China   Feng Lyu
Shanghai, China   Minglu Li
Waterloo, ON, Canada   Xuemin Shen

Contents

Acronyms

ACK	Acknowledgement
AP	Access point
B5G	Beyond 5G
BS	Base station
CAM	Cooperative awareness message
CCDF	Complementary cumulative distribution function
CCH	Control channel
CDF	Cumulative distribution function
CDMA	Code division multiple access
CSMA/CA	Carrier sense multiple access/collision avoidance
CTS	Clear to send
DCF	Distributed coordination function
DSRC	Dedicated short range communications
FCC	Federal Communications Commission
GPS	Global positioning system
ITS	Intelligent transportation system
LiDAR	Light detection and ranging
LoS	Line-of-sight
LTE	Long term evolution
MAC	Medium access control
MEC	Mobile edge computing
NLoS	None-line-of-sight
OBU	Onboard unit
OHS	One-hop set
PDF	Probability density function
PDR	Packet delivery ratio
PIL	Packet inter-loss
PIR	Packet inter-reception
QoS	Quality of service
RSU	Road side unit
RTS	Request to send

RX	Receiver
SAG	Space-air-ground
SAGVN	Space-air-ground integrated vehicular network
SCH	Service channel
SDMA	Space division multiple access
SDN	Software-defined networking
SUMO	Simulation of Urban MObility
TDMA	Time division multiple access
THS	Two-hop set
TPC	Transmission power control
TRC	Transmission rate control
TVWS	TV white spaces
TX	Transmitter
UAV	Unmanned aerial vehicles
URLLC	Ultra-reliability and low-latency communications
V2I	Vehicle-to-infrastructure
V2P	Vehicle-to-pedestrian
V2V	Vehicle-to-vehicle
V2X	Vehicle-to-everything
VANETs	Vehicular Ad hoc NETworks
WAVE	Wireless access in vehicular environments
WSMP	Wave short message protocol

Chapter 1
Introduction

Vehicular networks can be effective solutions to enhance road safety, via which vehicles can exchange cooperative awareness messages rapidly, contributing to better situation awareness and maneuvering cooperation. To well support road-safety applications, low-latency and reliable broadcast communications are required. However, with fast-changing network topologies, intermittent wireless links, and dynamic traffic densities in vehicular environments, it is challenging to achieve satisfying broadcasting performance, hereby expecting systematical and in-depth research. In this chapter, we first overview the vehicular network and then briefly introduce the V2X-enabled road-safety applications. Finally, we highlight the technical challenges of vehicular networking, and follow with our monograph organization with demonstrating the according contributions.

1.1 Vehicular Networks

With rapidly-growing urbanization, the research of Vehicular Ad hoc NETworks (VANETs) is fueled by two major social impetuses. The first one is the urgent need to assist in the transportation system by alleviating on-road problems, including crash accidents, traffic congestions, air pollution, i.e., improving road safety and transportation efficiency. The second one is the ever-increasing demand for mobile Internet access of passengers due to the thriving of mobile application industry, where vehicular users desire the mobile data services, such as online gaming, video streaming, mobile advertising, and so on [1, 2]. To meet the demands, VANETs integrate the technologies of wireless communication and informatics into the transportation system to facilitate the next generation Intelligent Transportation Systems (ITS), which can boost both the transportation performance and user driving experience [3, 4]. According to Cisco, building upon VANETS, new business models can be created ranging from building and serving vehicles, providing

© Springer Nature Switzerland AG 2020
F. Lyu et al., *Vehicular Networking for Road Safety*, Wireless Networks,
https://doi.org/10.1007/978-3-030-51229-3_1

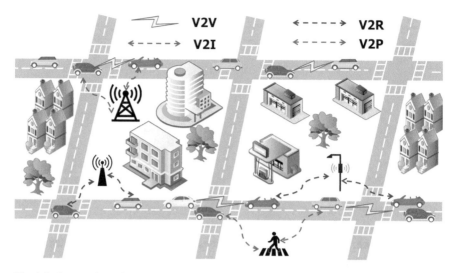

Fig. 1.1 An overview of vehicular networks

location-based services and cloud-based infotainment, to crash prevention with intelligent traffic management, which can create a win-win situation for society, vehicle users, and auto manufacturers [5].

As shown in Fig. 1.1, the vehicular network includes the communication paradigms of Vehicle-to-Vehicle (V2V), Vehicle-to-Road-Side-Unit (V2R), Vehicle-to-Infrastructure, Vehicle-to-Pedestrian (V2P), etc., broadly referred as Vehicle-to-Everything (V2X) communications. By enabling information exchanging among vehicles, communication infrastructure, and Internet, multifarious vehicular applications, such as road safety, real-time navigation, onboard entertainment, and self-driving, can be provided to drivers and passengers [6]. For instance, to enable road-safety applications [7], vehicles periodically broadcast cooperative messages of positional and kinematic information to one-hop neighbors, which can be beneficial to services of safety warning, collision avoidance, and speed/signal violation, etc. To guarantee the quality of service (QoS), the ultra-low delay and high reliability are usually required. On the contrary, to achieve comfort applications [8], such as file downloading, web browsing, and video streaming, vehicles need to fetch large-size contents from cloud/edge servers, where the high-bandwidth communications should be guaranteed.

Empowered by V2X communications, information generated by vehicles, onboard sensors, control system, or passengers, can be disseminated among communicating entities in proximity in real time. Generally, IEEE 802.11p based Dedicated Short Range Communication (DSRC) operating between the frequency of 5.700 and 5.925 GHz, can be adopted to support V2X communications, which is lightweight and easy to be implemented without any assistance of the built infrastructure. With direct communications among DSRC radios, the communication delay is acceptable, which can support a wide spectrum of delay-sensitive and road-safety applications, such as lane changing warning, intersection collision warning, and curve speed warning. However, the DSRC may fail to feed data-craving applications with high data rate requirements due to the spectrum scarcity problem [9]. To this end, one of the practical and seamless ways is to leverage the off-the-shelf 4G (i.e., Long Term Evolution (LTE)) or the emerging 5G cellular networks to provide V2I communications in order to satisfy high-rate Internet access [10, 11], since the cellular networks can provide large-area coverage with reasonable data rates. On the other hand, as cellular network services are with relatively high cost, users may prefer to use drive-thru WiFi or TVWS which is much cheaper, but users have to tolerate intermittent connectivity. Different spectrum resources have different access and management technologies, leading to prosperous vehicular researches in recent years. In this monograph, as we focus on enabling road-safety applications, we will concentrate on the DSRC networking techniques.

Fascinated by the visions of vehicular networks, various related activities have been initiated by the academia, industry, and government institutions around the world. An overview of the evolution of V2X technologies with both the pros and cons can be seen in the work [12]. The software-defined networking (SDN) enabled technologies to enhance future vehicular networks are envisioned in the article [13], in which the standards and standardization process are also presented. There are also other state-of-the-art surveys on VANETs in terms of performance analysis [14] and security [15].

1.2 Supporting Road-Safety Applications

Road safety has always been the first priority for daily commuters. According to the world health organization (WHO), more than 1.2 million and 50 million people worldwide are killed and injured due to collisions each year, respectively [16]. Being unresponsive to on-road emergencies is the major reason to most accidents, which poses the necessity of building active cooperative road-safety applications via VANETs [17]. Particularly, empowered by V2X communications, real-time environment information can be exchanged among neighboring vehicles rapidly via broadcasting road-safety beacons, i.e., cooperative awareness messages (CAMs),[1] that can facilitate various advanced road-safety applications with pre-sensing capabilities. To be specific, with V2V communications, each vehicle can periodically broadcast its status information including velocity, acceleration, heading direction, position, and turn signal status, to all one-hop neighbors. With such timely information, the application layer can support services such as pre-crash sensing, emergency electronic brake alert, blind spot warning, and cooperative forward collision avoidance [7]. On the other hand, with V2R communications,

---

[1] In this monograph, we use the word "beacon" and "CAM" interchangeably.

RSUs can periodically broadcast information such as the weather condition, speed limit, traffic signal status, road surface type, and current traffic condition, to all vehicles in proximity, whereby, services such as curve speed warning, stop sign violation, traffic signal violation, and among others, can be provided.

Compared with other sensors, such as camera, radar, and light detection and ranging (LiDAR), V2X communications can provide 360° situational awareness on road with offering more excellent sensing range, through-objects view functionality, and around-corner viewing capability. Besides, V2X communications are not affected or influenced by non-ideal weather conditions, such as heavy rain, fog, and harsh sunbeams, which can work robustly in real driving environments. Both advantages can collectively facilitate the extensive usage of V2X in future transportation systems, especially for road safety enhancement.

1.3 Networking Challenges

To well support road-safety applications, low-latency and reliable V2X communications are required. However, it is challenging to achieve the two goals under vehicular environments with fast-changing network topologies, intermittent wireless links, and dynamic traffic densities. The challenges can be summarized as follows.

- **The MAC Layer.** As the DSRC has scarce spectrum resources while the high-priority road-safety applications usually call for high-frequency broadcast, i.e., 10 Hz (every 100 ms), it poses enormous pressures on the medium resource management. Beside, as CAMs are related to road safety, minimizing the medium access delay and avoiding the medium access collision should be achieved simultaneously. However, due to the lack of a global central unit in vehicular environments, vehicles have to negotiate the medium access in a fully distributed way. Meanwhile, with the time-varying network topology, diverse spatial vehicle densities, and the hidden/exposed node problems, the MAC protocol has to work robustly, i.e., seamlessly adapting to dynamic communicating environments. At last, as there are various road-safety applications which may have different beaconing[2] rates, the MAC protocol should be scalable to enable their corporate negotiations.
- **The Link Layer.** The wireless link is unstable especially when vehicles moving fast, which can significantly affect the communication reliability. However, to characterize and understand the link behavior in urban environments is quite challenging. First, as urban environments are complex and highly dynamic, too many uncontrollable factors, such as time-varying traffic conditions, various types of roads, and all different surrounding trees and buildings, can affect V2V

---

[2]In this monograph, the words "beaconing" and "broadcasting" are interchangeable.

link performance. It is hard to separate the impact of each factor on the final DSRC link performance. Second, to conduct realistic studies on urban V2V communications, experiments should involve different traffic conditions, road types, and cover a sufficiently long time, which are labor-intensive and time-consuming. The lack of real-world trace is the hurdle of achieving efficient protocol design and precise model developing. Third, to thoroughly capture the link variation in the moving, various metrics should be comprehensively investigated. Performance analytics with single or limited metrics may not only give one-sided communication knowledge, but also confuse researchers and application designers without providing multi-perspective clues. In addition, to compensate for the link loss, an efficient relay scheme design is required, which should guarantee the reliability without wasting too much resource to congest the channel.

- **The Network Layer.** With the limited available V2X bandwidth, it is non-trivial to guarantee the road-safety demand for each vehicle, especially under dense-vehicle scenarios. On the one hand, if vehicles adopt aggressive beaconing rates, some vehicles may be sacrificed and have no required bandwidth to broadcast their moving status. On the other hand, if vehicles adopt moderate beaconing rates, the received moving status of neighboring vehicles may be out-of-date, resulting in delayed reactions to dangerous situations. In fact, moving vehicles are going to have different danger levels, calling for distinct beaconing rates to meet demands for road safety enhancement, and therefore the traditional congestion control approaches targeting at system throughput maximization are no longer effective in vehicular networks. In addition, there is normally no global central unit in vehicular environments, making it difficult to achieve the optimal beaconing scheme. Third, due to the high mobility of vehicles, the durations of V2X communications are very short. It is very important to minimize the communication overhead of a distributed beaconing scheme. Moreover, as the environment (e.g., the channel utility and the number of related vehicles) changes fast, such a distributed beaconing scheme should also react fast to keep the pace.

1.4   Aim of the Monograph

This monograph aims to investigate vehicular networking technologies for road safety enhancement. Specifically, we focus on DSRC technologies at the MAC, link, and network layers. Firstly, we review the state-of-the-art vehicular networking researches and organize a comprehensive survey to state our technical motivations per the MAC, link, and network layer. Then, we propose and design vehicular networking technologies at each corresponding layer to guarantee low-latency and reliable V2X communications for road safety enhancement.

In Chap. 3, to avoid medium access collisions caused by vehicular mobilities, we propose a mobility-aware TDMA-based MAC, named *MoMAC*, which can assign each vehicle a collision-avoidance time slot according to the underlying road topology and lane distribution on roads [18]. In *MoMAC*, different lanes on the same road segment and different road segments at intersections are associated with disjoint time slot sets, i.e., assigning vehicles that are bound to merge, with collision-free time slots. The merit of this design is that each vehicle can easily obtain the collision-avoidance time-slot assignment as long as the vehicle has a lane-level digital map and knows its current positional information of belonging to which road and which lane, which can be easily obtained by all the navigation systems. To achieve a common agreement about the time slot usage among neighboring vehicles, each vehicle is required to broadcast road-safety beacons together with the time slot occupying information of neighboring vehicles. By updating the time slot occupying information of two-hop neighbors (obtained indirectly from one-hop neighbors), vehicles can detect time slot collisions and access a vacant time slot in a fully distributed way. With coping with the mobility issue, *MoMAC* can significantly alleviate the medium access collisions and improve both the beacon transmission and reception rates.

As the existing vehicular MACs do not consider the situation that vehicles have diverse beaconing rates to support various road-safety applications, and such inflexible design may suffer from a scalability issue in terms of channel resource management, in Chap. 4, we propose a novel time Slot-Sharing MAC, named *SS-MAC*, to support diverse beaconing rates of vehicles [19]. Particularly, we first introduce a circular recording queue to perceive the time slot occupying status in real time. We then design a distributed time slot sharing (DTSS) approach and random index first fit (RIFF) algorithm, to efficiently share the time slot and conduct the online vehicle-slot matching, respectively. By supporting vehicles with different beaconing rates to share the same time slot, *SS-MAC* can work flexibly to significantly improve the medium resource utilization. Note that, our proposed *MoMAC* and *SS-MAC* can work collectively to provide collision-free/reliable, scalable, and efficient medium access for moving and distributed vehicles.

With limited literature available, there is a lack of understanding about how IEEE 802.11p based DSRC performs for V2V communications in urban environments. In Chap. 5, we investigate the vehicular link layer performance, which is of paramount importance for reliable information exchanging considering intermittent wireless links [20]. Particularly, we implement a V2V communication testbed with two experimental vehicles, each equipped with an off-the-shelf IEEE 802.11p-compatible onboard unit. We conduct intensive data analytics on V2V communication performance, based on the field measurement data collected from different environments in Shanghai city, and obtain several key insights as follows. First, among many context factors, non-line-of-sight (NLoS) link condition is the major factor degrading V2V performance. Second, both line-of-sight (LoS) and NLoS durations follow power law distributions, which implies that the probability of having long LoS/NLoS conditions can be relatively high. Third, the packet inter-reception (PIR) time distribution follows an exponential distribution in LoS conditions but a power law in NLoS conditions. In contrast, the packet inter-loss (PIL) time distribution in LoS condition follows a power law but an exponential in NLoS condition. This means that consecutive packet reception failures can rarely appear when in LoS conditions but can constantly appear when in NLoS conditions. Fourth, the overall PIR time distribution is a mix of exponential distribution and power law distribution. The presented results can provide solid ground to validate models, tune VANET simulators, and improve communication strategies.

With characterizing the link communication performance, in Chap. 6, we investigate link-aware beaconing scheme design to enhance the communication reliability [20]. Particularly, based on the observation that among many types of contextual information, NLoS link condition is the key factor of V2V performance degradation, we propose a link-aware reliable beaconing scheme, named *CoBe* (i.e., Cooperative Beaconing), to enhance the broadcast reliability for road-safety applications. *CoBe* is a fully distributed scheme, in which a vehicle first detects the link condition with each of its neighbors by machine learning algorithms, then exchanges such link condition information with its neighbors, and finally selects the minimal number of helper vehicles to rebroadcast its beacons to those neighbors in bad link condition. To analyze and evaluate the performance of *CoBe* theoretically, we devise a two-state Markov chain model to mimic beaconing behaviors under LoS/NLoS conditions. In addition to *CoBe*, we also present a case study of efficient unicasting scheme for non-road-safety applications. With a deep understanding of the link features, our proposed link-aware scheme design can achieve both the beaconing reliability and efficient resource utilization since only essential rebroadcasts are triggered to take necessary actions.

After enhancing the beaconing performance at both the MAC layer and link layer, in Chap. 7, we turn to the network layer performance [21]. Particularly, under dynamic traffic conditions, especially for dense-vehicle scenarios, the naive beaconing scheme where vehicles broadcast beacons at a fixed rate with a fixed transmission power can cause severe channel congestion and degrade the beaconing reliability. To this end, by considering the kinematic status and beaconing rate together, we study the rear-end collision risk and define a danger coefficient ρ to capture the danger threat of each vehicle being in the rear-end collision. Based on individually estimated ρ, we propose a fully distributed adaptive beacon control scheme, named *ABC*, which makes each vehicle actively adopt a minimal but sufficient beaconing rate to avoid the rear-end collision in dense scenarios. With *ABC*, vehicles can broadcast at the maximum beaconing rate when the channel medium resource is enough and meanwhile keep identifying whether the channel is congested. Once a congestion event is detected, an NP-hard distributed beacon rate adaptation (DBRA) problem is solved with a greedy heuristic algorithm, in which a vehicle with a higher ρ is assigned with a higher beaconing rate while keeping the total required beaconing demand lower than the channel capacity. By adopting our proposed *ABC*, vehicles can adapt beaconing rates in accordance with the road safety demand with an acceptable communication overhead, and the beaconing reliability can be guaranteed even under high-dense vehicle scenarios.

Finally, we organize a summary for this monograph and discuss potential future research directions of vehicular networks, including space-air-ground integrated vehicular network, software-defined networking, and mobile edge computing.

References

1. N. Lu, N. Cheng, N. Zhang, X. Shen, J.W. Mark, Connected vehicles: solutions and challenges. IEEE Internet Things J. **1**(4), 289–299 (2014)
2. J.E. Siegel, D.C. Erb, S.E. Sarma, A survey of the connected vehicle landscape, architectures, enabling technologies, applications, and development areas. IEEE Trans. Intell. Transp. Syst. **19**(8), 2391–2406 (2018)
3. X. Cheng, R. Zhang, L. Yang, Wireless toward the era of intelligent vehicles. IEEE Internet Things J. **6**(1), 188–202 (2019)
4. N. Cheng, F. Lyu, J. Chen, W. Xu, H. Zhou, S. Zhang, X. Shen, Big data driven vehicular networks. IEEE Netw. **32**(6), 160–167 (2018)
5. Cisco, The internet of cars. https://www.cisco.com/c/en/us/solutions/industries/transportation/passenger.html
6. F. Lyu, N. Cheng, H. Zhu, H. Zhou, W. Xu, M. Li, X. Shen, Intelligent context-aware communication paradigm design for IoVs based on data analytics. IEEE Netw. **32**(6), 74–82 (2018)
7. CAMP Vehicle Safety Communications Consortium and Others, Vehicle safety communications project: task 3 final report: identify intelligent vehicle safety applications enabled by DSRC. *National Highway Traffic Safety Administration, US Department of Transportation, Washington DC*, March 2005
8. H. Zhou, N. Cheng, N. Lu, L. Gui, D. Zhang, Q. Yu, F. Bai, X. Shen, WhiteFi infostation: engineering vehicular media streaming with geolocation database. IEEE J. Sel. Areas Commun. **34**(8), 2260–2274 (2016)
9. H. Zhou, N. Cheng, Q. Yu, X. Shen, D. Shan, F. Bai, Toward multi-radio vehicular data piping for dynamic DSRC/TVWS spectrum sharing. IEEE J. Sel. Areas Commun. **34**(10), 2575–2588 (2016)
10. S. Chen, J. Hu, Y. Shi, Y. Peng, J. Fang, R. Zhao, L. Zhao, Vehicle-to-everything (V2X) services supported by LTE-based systems and 5G. IEEE Commun. Stand. Mag. **1**(2), 70–76 (2017)
11. S. Chen, J. Hu, Y. Shi, L. Zhao, LTE-V: a TD-LTE-based V2X solution for future vehicular network. IEEE Internet Things J. **3**(6), 997–1005 (2016)
12. H. Zhou, W. Xu, J. Chen, W. Wang, Evolutionary V2X technologies toward the internet of vehicles: challenges and opportunities. Proc. IEEE **108**(2), 308–323 (2020)
13. W. Zhuang, Q. Ye, F. Lyu, N. Cheng, J. Ren, SDN/NFV-empowered future IoV with enhanced communication, computing, and caching. Proc. IEEE **108**(2), 274–291 (2020)
14. Y. Ni, L. Cai, J. He, A. Vinel, Y. Li, H. Mosavat-Jahromi, J. Pan, Toward reliable and scalable internet of vehicles: performance analysis and resource management. Proc. IEEE **108**(2), 324–340 (2020)
15. K. Ren, Q. Wang, C. Wang, Z. Qin, X. Lin, The security of autonomous driving: threats, defenses, and future directions. Proc. IEEE **108**(2), 357–372 (2020)
16. World Health Organization (WHO), World report on road traffic injury prevention. https://www.who.int/publications-detail/world-report-on-road-traffic-injury-prevention
17. H. Peng, L. Liang, X. Shen, G.Y. Li, Vehicular communications: a network layer perspective. IEEE Trans. Veh. Technol. **68**(2), 1064–1078 (2019)
18. F. Lyu, H. Zhu, H. Zhou, L. Qian, W. Xu, M. Li, X. Shen, MoMAC: mobility-aware and collision-avoidance MAC for safety applications in VANETs. IEEE Trans. Veh. Technol. **67**(11), 10590–10602 (2018)
19. F. Lyu, H. Zhu, H. Zhou, W. Xu, N. Zhang, M. Li, X. Shen, SS-MAC: a novel time slot-sharing MAC for safety messages broadcasting in VANETs. IEEE Trans. Veh. Technol. **67**(4), 3586–3597 (2018)

20. F. Lyu, H. Zhu, N. Cheng, H. Zhou, W. Xu, M. Li, X. Shen, Characterizing urban vehicle-to-vehicle communications for reliable safety applications. IEEE Trans. Intell. Transp. Syst. 1–17, Early Access (2019). https://doi.org/10.1109/TITS.2019.2920813
21. F. Lyu, N. Cheng, H. Zhu, H. Zhou, W. Xu, M. Li, X. Shen, Towards rear-end collision avoidance: adaptive beaconing for connected vehicles. IEEE Trans. Intell. Transp. Syst. 1–16, Early Access (2020). https://doi.org/10.1109/TITS.2020.2966586

Chapter 2
Vehicular Networking Techniques for Road-Safety Applications

As the performance of road-safety applications depends on the communication quality in terms of delay and reliability, which can be affected at the MAC, link, and network layers, in this chapter, we provide a comprehensive survey of vehicular networking techniques for road-safety applications. Particularly, we review the state-of-the-art related researches in three sections: (1) MAC design; (2) Link quality characterization and enhancement; and (3) network congestion control.

2.1 MAC Design

MAC protocol is crucial for broadcasting performance as it can essentially affect both the access delay and access collision. We review the vehicular MAC protocols in two categories, i.e., contention-based MAC protocols and contention-free MAC protocols [1–4].

2.1.1 Contention-Based MAC Protocols

For the contention-based MAC protocol, such as IEEE 802.11p, it is a customized standard as an amendment to the existing IEEE 802.11a-2007 or Wi-Fi standard [5], and has been dedicated by the Federal Communications Commission (FCC) as the MAC layer standard for DSRC-enabled V2X communications in the United States. Although the standards of IEEE 802.11 have been widely implemented and adopted in networking systems, IEEE 802.11p may fail to guarantee efficient broadcasting communications in vehicular networks due to the following reasons. First, as the basic mechanism of IEEE 802.11p is the same as the IEEE 802.11 standards that adopt the distributed coordination function (DCF), the mechanism

© Springer Nature Switzerland AG 2020
F. Lyu et al., *Vehicular Networking for Road Safety*, Wireless Networks,
https://doi.org/10.1007/978-3-030-51229-3_2

works with the carrier sense multiple access/collision avoidance (CSMA/CA). Within the CSMA/CA, the contention process is performed among communicating nodes. Particularly, if a node wants to access the medium, it will first sense the channel using status; if the channel is perceived to be free without accessing, the node can occupy and access the medium immediately, otherwise, the node has to wait and perform the random back-off procedure. This contention process can result in unbounded delays when there are too many contending nodes [6–9], which cannot guarantee the real-time requirement of road-safety applications. Second, in the broadcast mode of IEEE 802.11p protocol, to expedite the real-time response, packets of request to send (RTS) /clear to send (CTS) /acknowledgement (ACK) are removed, where the hidden terminal problem can arise. Specifically, many studies have analyzed the broadcasting performance of 802.11p DCF-based mechanism in terms of delay and reliability [10], and they have disclosed that the MAC protocol of IEEE 802.11p generally has issues of unbounded delay and serious channel congestion under high-density environments [11, 12]. For instance, in the work [13], the authors have confirmed that even though the number of collisions can be reduced with dynamically adjusting the contention window of IEEE 802.11p protocol, the packet reception probability can hardly reach 90% due to the randomness feature of underlying CSMA-based scheme. In addition, according to the observations in the work [14], the beacon delays normally last 200 ms but can be larger than 500 ms and sometimes reach above 1 s under a dense environment. To this end, it is widely recognized that the contention-based MAC protocols are not suitable for road-safety applications with stringent real-time requirements.

2.1.2 Contention-Free MAC Protocols

Unlike contention-based MAC protocols, contention-free MAC protocols work under a channel usage agreement among communicating nodes prior to data transmission. Several types of contention-free MAC protocols have been proposed, including Space Division Multiple Access (SDMA)-based, Code Division Multiple Access (CDMA)-based, and Time Division Multiple Access (TDMA)-based MAC protocols.

SDMA-Based MAC Protocols In SDMA-based MAC protocols, vehicles acquire medium resources according to the underlying physical locations [15, 16]. Those MAC protocols usually work with the following three steps: (1) discrete processing: target roads are divided into several small and independent road units; (2) mapping function: unique time slot sets are associated with each road unit; (3) assignment rules: with the positional information of which road unit that vehicles move on, the system determines which time slots are available for vehicles to access. SDMA-based MAC protocols require precise positional information of vehicles, while the missing or inaccuracy of such information may result in frequent transmission collisions. In addition, it is difficult to achieve efficient resource allocation and utilization. Particularly, when the traffic density is low or uneven on some road segments, the medium resources are likely to be wasted. On the other hand, when the traffic density is high, the medium resources may be insufficient, and then it may fail to ensure the fairness of medium access for all the vehicles [17].

CDMA-Based MAC Protocols In CDMA-based MAC protocols, channel resources are distinguished with different pseudo random codes [18, 19]. The receivers usually adopt the same pseudo noise (PN) code to demodulate the received signals. As this mechanism is robust to combat with the interference and noise with bandwidth utilization enhancement, many CDMA-based MAC protocols have been proposed for vehicular message dissemination. However, in vehicular networks with high-density vehicles, the required PN codes to differentiate each vehicle can be relatively long, inflicting significant communication overhead. Besides, to demodulate packets, receivers have to save the PN codes of each sender, which is unrealistic since it is impractical for vehicles to save all PN codes of neighbors within the communication range.

TDMA-Based MAC Protocols Recently, TDMA-based MAC protocols have been proposed to support broadcast communications for road-safety applications [2, 20]. In TDMA-based MAC protocols, time is partitioned into frames, each containing a constant number of equal-length time slots which are synchronized among communicating vehicles. Each vehicle can be granted to access the channel once in each frame with occupying a unique time slot. The time-slotted channel can ably guarantee the stringent delay requirement of road-safety applications. In addition, during broadcast, vehicles also include the status of time slot usage of one-hop neighbors in each beacon. In doing so, vehicles can perceive the up-to-date time slot usage of two-hop neighbors, based on which vehicles can acquire distinct time slots with each other, and be able to detect access collisions and avoid the hidden/exposed terminal problems without RTS/CTS/ACK schemes [21].

There have been some studies on TDMA-based MAC design in vehicular networks. Hadded et al. [22] organized a survey on TDMA-based MAC protocols, in which they first discussed the features of vehicular networks and the stringent delay requirement of road-safety applications, and then justified the motivations of adopting the TDMA-based MAC protocols. After that, they provided an overview of TDMA-based MAC protocols proposed in the literature with summarizing and comparing their characteristics, benefits, and limitations. Centralized TDMA-based MAC protocols have been proposed in studies [23–25], in which a central controller usually takes the responsibility to assign and allocate time slots. Specifically, in the work [23], Zhang et al. utilized an RSU as a centralized controller to collect the channel state information and vehicle individual information. With the information, the controller can calculate the respective scheduling weight factors, and then make scheduling decisions. However, the protocol requires a large number of RSUs as controllers, which can limit the network scale and is unable to work when there is no RSUs. In another two protocols [24, 25], several cluster heads are chosen as central controllers, where vehicles are partitioned into several clusters, each choosing a head vehicle as the controller. However, the mechanism is impractical in vehicular networks due to the high mobilities of vehicles, which poses great challenges in cluster forming and cluster head selection when the network topologies vary dramatically with time. To this end, many studies focus on distributed TDMA-based MAC design, in which each vehicle negotiates the time slot usage in a distributed manner.

The main challenge for distributed TDMA-based MAC protocol is how to coordinate the time slot usage efficiently among vehicles when there is no centralized control and vehicles have no global network information. The protocol named ADHOC MAC was proposed in the work [26], in which the wireless communication channel is represented by a slotted/framed structure, and shared by vehicles with the well-known Reservation ALOHA (R-ALOHA) protocol. However, those researches do not consider the mobility impacts on time slot allocation, which can suffer from significant performance degradation in real-driving scenarios. Considering the transmission collisions caused by vehicular movements in opposite directions, Omar et al. proposed a protocol named VeMAC [2] for reliable broadcast communications in VANETs. VeMAC assigns disjoint time slot sets to vehicles moving in opposite directions to reduce transmission collisions. Based on VeMAC, ATSA (Adaptive TDMA Slot Assignment) MAC [27] was proposed to enhance VeMAC performance when the vehicle densities in opposite directions are uneven, in which the frame length is dynamically shortened or doubled with the algorithm based on binary tree. For both two protocols, the considered moving conditions are assumed to be stable, where the speed, moving direction, and distance among vehicles are constant. This strong assumption cannot stand in practical VANETs. On the other hand, Jiang et al. [28] proposed PTMAC, i.e., a prediction-based TDMA-based MAC protocol, to reduce access collisions for broadcast communications. In PTMAC, two-way traffic and four-way intersections are considered. With collecting speed, position, and moving direction information from neighbors, PTMAC first predicts the possibility of encountering collisions and then tries to decrease the potential collisions. However, PTMAC needs to seek intermediate vehicles to conduct potential collision detection and potential collision elimination, causing extra delay with communication overhead. In addition, PTMAC relies on intermediate vehicles for coordination, constraining its usage when there is no intermediate vehicle in the environment.

Due to the merits of TDMA-based MAC protocols in supporting broadcast communications, we will adopt the TDMA-based mechanism to cast our MAC design, in which we will systematically investigate the mobility impacts and figure out mobility-aware protocol design to provide collision-free broadcast communications. In addition, in current TDMA-based MAC protocols, each vehicle is configured with the same broadcasting rate, where the situation of diverse beaconing rates of vehicles is not considered. Such inflexible design would suffer from a serious scalability issue in terms of medium resource management when road-safety applications call for different broadcasting rates. We will also investigate the techniques of efficient and scalable MAC design for road-safety applications.

2.2 Link Quality Characterization and Enhancement

For vehicular communications, the link quality is significant since the wireless channel is vulnerable to the link condition in terms of communicating distance, multipath fading (caused by both mobile (vehicles) and stationary (e.g., trees and buildings) scatterers), as well as shadowing (caused by in-channel obstacles) [29, 30]. We review the related work at link layer in two categories, i.e., link quality characterization and relay scheme design for reliability enhancement.

2.2.1 Link Quality Characterization

Measurement-Based Studies There have been some measurement-based studies on characterizing DSRC link performance in the literature. For instance, to investigate the impacts of propagation environment, communicating distance, mobility speed, and transmission power on the final V2V link performance, Bai et al. [31] conducted an empirical study on packet delivery ratio (PDR), referring to the probability of the receiver successfully receiving a packet transmitted from the sender, in different scenarios. In addition, they analyzed the spatio-temporal variation and symmetric correlation of PDR performance, and depicted the PDR variation with different parameters. However, using the single metric of PDR is not sufficient to characterize the underlying features of intermittent V2V link quality, especially under complicated urban environments. In addition to PDR, Martelli et al. studied the distribution of packet inter-reception (PIR) time, referring to the interval of time elapsed between two successfully received packets, and investigated the relationship between PDR and other environmental factors such as speed and communicating distance [32]. They claimed that the PIR time distribution follows a power law distribution, and the PIR time is almost independent of speed and communicating distance and is loosely correlated with PDR. The observations are of significant importance for vehicular networking, but these characteristics may change under different channel conditions. In the paper [33], based on real-world experimental data, the authors characterized the application-level communication reliability of DSRC for road-safety applications and concluded that the V2V communication reliability of DSRC is adequate since packet reception failures do not occur in bursts, the analysis granularity of which is too coarse to guarantee extremely reliable communications for road-safety applications. J. Gozalvez et al. conducted extensive campaigns of field-testing to investigate the quality of IEEE 802.11p enabled V2I communications in urban scenarios [34]. They reported that the urban environment, street layout, traffic density, trees, terrain elevation, and in-channel large vehicles have impacts on V2I communications, which should be taken into account when deploying and configuring urban RSUs. However, all these characteristics can vary significantly when in different channel conditions in terms of line-of-sight (LoS) and non-line-of-sight (NLoS) conditions, and aggregating measurements together to draw conclusions may bias from the ground truth.

There are also some measurements on physical layer of DSRC channels, where the coherence time, path loss, and doppler spectrum are analyzed [35, 36]. For example, in the work [35], wireless channel impairments on DSRC performance are measured and analyzed, and the results suggest that although the proposed DSRC standard may account for doppler and delay spreads for vehicular channels, many packets may face high error rates with time-varying channel qualities. In the work [36], the authors adopted the receivers with differential global positioning systems to conduct dynamic measurements on how coherence time, large-scale path loss, and doppler spectrum vary with vehicle location and communicating distance. However, all these findings could be quite different when vehicles move or the underlying environment changes. In this monograph, we do not put much emphasis on physical layer features as they vary dramatically in the moving, which can be hardly and qualitatively characterized with determined patterns.

Regarding LoS and NLoS effects, Meireles et al. [37] conducted the experimental study and confirmed that the channel quality can be deeply affected by LoS and NLoS conditions. Particularly, under several scenarios, they collected the PDR as well as the received signal power information. With the information, they then quantified the obstruction impact, where they concluded that NLoS conditions can effectively halve the communication range within which 90% of communications can be successful. This insight is valuable, but they collected data with fixing vehicles and obstructions, which may bias from the results of moving scenarios. For LoS and NLoS modeling, M. Boban et al. designed a geometry-based efficient propagation model for V2V communication, in which LoS and NLoS conditions are taken into account [38]. In our research, we do not model the LoS/NLoS channel but concentrate on the LoS and NLoS interactions in the moving and characterize V2V performance under two distinct channel conditions.

2.2.2 Relay Scheme Design

With link quality measurement, it is widely known that the vehicular wireless channel may undergo the unreliability issue in the moving, especially when encountering NLoS conditions. For road-safety applications, as ultra reliable communication is required, some relay schemes are proposed to enhance the link reliability. In this subsection, we review the techniques of relay scheme design.

Receiver-Oriented Schemes In the receiver-oriented schemes, all vehicles that have received the beacon, will contend for being relays under a predefined mechanism, such as probability-based or waiting-time-based mechanisms. Wisitpongphan et al. proposed three probability-based relaying schemes, i.e., slotted p-persistence, slotted 1-persistence and weighted p-persistence [39]. They work with a similar manner, in which if one vehicle receives a beacon from the sender for the first time, it will rebroadcast the packet with the probability p; otherwise, it will drop the packet. The difference between each scheme is how to set the value of p. Particularly, in 1-persistence and p-persistence scheme, the value of p is set to 1 and a pre-determined value, respectively. Differently, in the weighted p-persistence scheme, the probability p is set to $\frac{D}{R}$ where D is the communicating distance between the sender and receiver, and R is the average communication range. For waiting-time-based schemes, a scheme is proposed in the work [40], in which each candidate relay vehicle determines the waiting time based on the distance d to the source vehicle, and the vehicle with a larger d will be associated with a smaller waiting time. Yang et al. proposed a scheme named PAB (i.e., position-based adaptive broadcast), to calculate the waiting time based on the position and speed of source and candidate relay vehicles [41]. Similar approaches are also used in the schemes of Opportunistic broadCast (OppCast) [42] and Urban Vehicular BroadCAST (UV-CAST) [43]. In addition, YOO et al. proposed the scheme of ROFF (i.e., RObust and Fast Forwarding), in which they assign the waiting time for each candidate relay vehicle based on their forwarding priority, and the priority is obtained via the empty space of beacon message dissemination between vehicles [44]. In the protocol of ABSM (Acknowledged Broadcast from Static to highly Mobile), upon receiving a beacon, instead of retransmitting it immediately, the vehicle will wait to check if retransmissions from other vehicles would cover its whole neighborhood [45].

Aforementioned relay schemes are easy to be implemented since they can be run locally without complicated negotiation. However, as communication contexts are not considered, they are unable to react to dynamic environments, which could result in broadcast storm problem and cause channel resource wasting with broadcasting too many duplicated beacons.

Sender-Oriented Schemes   On the contrary, in the sender-oriented relay schemes, the source vehicle will explicitly select potential vehicles as relays, and those vehicles with successfully receiving the beacon can be listed as potential relays. As the sender-oriented schemes limit the number of contending relays proactively, efficient channel resource utilization can be guaranteed. However, they require additional information of vehicles to assist in the relay selection, and thus the system performance deeply depends on the quality of input information, i.e., what kind of information is input and how the information can be achieved in real driving environments. For example, Rehman et al. proposed the scheme of BDSC (i.e., Bi-directional Stable Communication), in which they investigated the relations between the estimated link quality and communicating distance [46]. After that, the relay vehicles can be selected based on the quantitative representation of link quality. The performance of BDSC is only evaluated with theoretical analysis, but in practice, the communicating distance is not the only factor that can affect link performance. Boban et al. indicated that tall vehicles with elevated antenna positions can improve communication performance since the communication links are more likely to encounter LoS conditions, inspired by which they proposed the scheme of TVR (i.e., Tall Vehicle Relaying) with distinguishing between short and tall vehicles and choosing tall vehicles as next-hop relays [47]. The scheme cannot work robustly when there is no tall vehicles in the environment, or there are slopes (i.e., the road is not flat) between communicating vehicles.

Nevertheless, there is no statistical study on the impact of channel conditions in terms of LoS and NLoS on the DSRC performance in urban environments and how these two conditions interact in the moving, which will be our technical emphasis in this monograph. With link quality investigation, we will also figure out the context-aware relay scheme design to achieve both the efficient resource utilization and satisfying reliability performance.

2.3 Network Congestion Control

Network layer activities can also significantly affect the broadcasting reliability, where the high beaconing rates of vehicles may cause severe channel congestion, especially under high-dense environments [48–51]. In the literature, congestion control approaches generally fall into three categories, i.e., transmit data rate control (TDC), transmit power control (TPC), and transmit message rate control (TRC). For road-safety applications, as the beacon size is small, TDC-based approaches are not efficient in general and will not be considered in this monograph.

2.3.1 Transmit Power Control (TPC)

There have been several TPC-based approaches for vehicular networks. For example, the D-FPAV scheme was proposed in [52], targeting at guaranteeing max-min fairness among vehicles within the channel load capacity. The D-FPAV algorithm is devised to calculate the allowed maximum Tx powers for vehicles. As the original D-FPAV design has a defect due to the heavy packet overhead, Mittag et al. devised an upgraded version, which can reduce two orders of magnitude in overhead [53]. Shah et al. proposed a TPC-based approach, named AC3, which can allow vehicles to determine their transmission powers automatically in accordance with the local channel congestion conditions [54]. In AC3, the authors defined a notion to indicate the vehicle marginal contribution, which is aggregated from vehicles to form a potential channel congestion; based on the cooperative game theory, the vehicle with the highest marginal contribution should reduce the most transmission power. On the other hand, joint rate and power control of broadcast communication is also investigated. For instance, Egea-Lopez et al. proposed that vehicles can deliver packets with different transmission power levels, and each level is associated with a distinct beaconing rate [55]. The rate selection problem is then modeled as the network utility maximization (NUM) problem. Complying with the allowed maximum beaconing load constraint, the objective is to maximize the number of beacons delivered at each transmit power level. In the work [56], to comply with the tracking error demand, the required minimum transmission rates of vehicles are first calculated, the transmission powers of vehicles are then enlarged until the channel busy ratio reaches over a pre-defined threshold. For these proactive approaches, i.e., adjusting transmission power proactively to avoid future potential channel congestions, precise models are required to carry out accurate prediction of future channel load, vehicle density, etc. However, with high-mobility vehicles and time-varying traffic in dynamic vehicular environments, the model quality is usually difficult to be guaranteed, constraining the effectiveness of these approaches. Specifically, the recent proposal [57] has studied TPC-based approaches and pointed out that the system performance can be significantly affected by the qualities of the transmission and prediction models, leading to a serious instability issue. In addition, previous studies [58, 59] also have concluded that the message rate control is the most efficient method to reach the stable stage. Therefore, we focus on TRC-based techniques to cast our congestion control scheme design.

2.3.2   Transmit Message Rate Control (TRC)

In regard to TRC-based approaches, LIMERIC [60] and PULSAR [61] are two highly cited researches, which have many similar techniques. Particularly, in LIMERIC [60], with up-to-date feedback of beaconing rate in use from neighbors, a linear control algorithm is designed, while in PULSAR [61], based on the binary feedback (congested or not) from two-hop neighbors, the authors devised an iteration algorithm named AIMD (i.e., additive increase multiplicative decrease) to adjust beaconing rates for vehicles. However, both pieces of work do not consider the driving context, i.e., all vehicles adapting to the same beaconing rate equally. Although such equal-fairness resource allocation can achieve the maximum system throughput, it cannot guarantee the possible best road-safety benefit for the transportation system. There are two recent proposals [62, 63] on the vehicular network congestion control, both of which model a Network Utility Maximization (NUM) problem for beaconing rate adaptation. Specifically, targeting at maximizing the system throughput, Egea-Lopez et al. defined a notion of "fairness" for beacon rate adaptation, and then proposed a algorithm named FABRIC to solve the dual of the NUM problem, in which the scaled gradient projection scheme is adopted [62]. Likewise, built upon a slotted p-persistent broadcast MAC, Zhang et al. also formulated a NUM problem for beacon rate adaptation, in which the velocity and relative position of vehicles are considered [63]. In both proposals, network utility is optimized but vehicle road-safety demands are not preferentially treated. The same issue also exists in the scheme of DBCC (i.e., distributed beacon congestion control) [64], in which beaconing medium resources are allocated according to vehicle link qualities rather than their driving contexts. In the work [65], each vehicle requests for the beaconing rate. The RSU as a centralized controller will allocate channel resources in accordance with the requests of vehicles, which is formulated as an optimization problem and then transformed into a maximum weighted independent set problem. As the scheme requires a centralized controller and takes time to converge to the optimal result, it may fail to satisfy the distributed and real-time requirements of broadcasting communications.

In this monograph, we will investigate the network-layer techniques of safety-aware and distributed beacon congestion control, which is rarely seen in the literature but essential for reliable road-safety applications.

2.4 Summary

In this chapter, we have surveyed state-of-the-art vehicular networking techniques for road-safety applications, where we have reviewed the related researches at the MAC layer, link layer, and network layer, respectively. Also, we have stated the corresponding technical motivations of this monograph in comparison with the current literature.

References

1. W. Xu, H. Zhou, N. Cheng, F. Lyu, W. Shi, J. Chen, X. Shen, Internet of vehicles in big data era. IEEE/CAA J. Autom. Sinica **5**(1), 19–35 (2018)
2. H.A. Omar, W. Zhuang, L. Li, VeMAC: a TDMA-based MAC protocol for reliable broadcast in VANETs. IEEE Trans. Mobile Comput. **12**(9), 1724–1736 (2013)
3. F. Lyu, H. Zhu, H. Zhou, W. Xu, N. Zhang, M. Li, X. Shen, SS-MAC: a novel time slot-sharing MAC for safety messages broadcasting in VANETs. IEEE Trans. Veh. Technol. **67**(4), 3586–3597 (2018)
4. N. Cheng, F. Lyu, J. Chen, W. Xu, H. Zhou, S. Zhang, X. Shen, Big data driven vehicular networks. IEEE Netw. **32**(6), 160–167 (2018)
5. IEEE Std 802.11-2007 (Revision of IEEE Std. 802.11-1999), Standard for Information Technology-Telecommunications and Information Exchange between Systems-Local and Metropolitan Area Networks-Specific Requirements - Part 11: Wireless LAN Medium Access Control (MAC) and Physical Layer (PHY) Specifications (2007), pp. 1–1184
6. IEEE Std 802.11p-2010, Standard for Information Technology-Telecommunications and Information Exchange between Systems-Local and Metropolitan Area Networks-Specific Requirements Part 11: Wireless LAN Medium Access Control (MAC) and Physical Layer (PHY) Specifications Amendment 6: Wireless Access in Vehicular Environments (2010), pp. 1–51
7. S. Subramanian, M. Werner, S. Liu, J. Jose, R. Lupoaie, X. Wu, Congestion control for vehicular safety: Synchronous and asynchronous MAC algorithms, in *Proceedings of the Ninth ACM International Workshop on Vehicular Inter-networking, Systems, and Applications* (ACM, New York, 2012), pp. 63–72
8. Q. Ye, W. Zhuang, L. Li, P. Vigneron, Traffic-load-adaptive medium access control for fully connected mobile ad hoc networks. IEEE Trans. Veh. Technol. **65**(11), 9358–9371 (2016)
9. H. Zhou, W. Xu, J. Chen, W. Wang, Evolutionary V2X technologies toward the internet of vehicles: Challenges and opportunities. Proc. IEEE **108**(2), 308–323 (2020)
10. H. Peng, D. Li, K. Abboud, H. Zhou, H. Zhao, W. Zhuang, X. Shen, Performance analysis of IEEE 802.11p DCF for multiplatooning communications with autonomous vehicles. IEEE Trans. Veh. Technol. **66**(3), 2485–2498 (2017)

11. Q. Chen, D. Jiang, L. Delgrossi, IEEE 1609.4 DSRC multi-channel operations and its impli-cations on vehicle safety communications, in *2009 IEEE Vehicular Networking Conference (VNC)* (2009), pp. 1–8

12. M.I. Hassan, H.L. Vu, T. Sakurai, Performance analysis of the IEEE 802.11 MAC protocol for DSRC safety applications. IEEE Trans. Veh. Technol. **60**(8), 3882–3896 (2011)

13. Y. Mertens, M. Wellens, P. Mahonen, Simulation-based performance evaluation of enhanced broadcast schemes for IEEE 802.11-based vehicular networks, in *VTC Spring 2008 - IEEE Vehicular Technology Conference* (2008), pp. 3042–3046

14. R. Reinders, M. van Eenennaam, G. Karagiannis, G. Heijenk, Contention window analysis for beaconing in VANETs, in *2011 7th International Wireless Communications and Mobile Computing Conference* (2011), pp. 1481–1487

15. J.J. Blum, A. Eskandarian, A reliable link-layer protocol for robust and scalable intervehicle communications. IEEE Trans. Intel. Transport. Syst. **8**(1), 4–13 (2007)

16. S.V. Bana, P. Varaiya, Space division multiple access (SDMA) for robust ad hoc vehicle communication networks, in *Proceedings of IEEE Intelligent Transportation Systems* (2001), pp. 962–967

17. F. Lyu, J. Ren, N. Cheng, P. Yang, M. Li, Y. Zhang, X. Shen, LEAD: Large-scale edge cache deployment based on spatio-temporal WiFi traffic statistics. IEEE Trans. Mob. Comput. 1–16 (2020). https://doi.org/10.1109/TMC.2020.2984261

18. F. Watanabe, M. Fujii, M. Itami, K. Itoh, An analysis of incident information transmission per-formance using MCS/CDMA scheme, in *Proceedings of IEEE Intelligent Vehicles Symposium* (2005), pp. 249–254

19. T. Inoue, H. Nakata, M. Itami, K. Itoh, An analysis of incident information transmission performance using an IVC system that assigns PN codes to the locations on the road, in *Proceedings of IEEE Intelligent Vehicles Symposium* (2004), pp. 115–120

20. F. Lyu, H. Zhu, H. Zhou, L. Qian, W. Xu, M. Li, X. Shen, MoMAC: mobility-aware and collision-avoidance MAC for safety applications in VANETs. IEEE Trans. Veh. Tech-nol. **67**(11), 10590–10602 (2018)

21. F. Lyu, J. Ren, P. Yang, N. Cheng, W. Tang, Y. Zhang, X. Shen, Fine-Grained TDMA MAC design toward ultra-reliable broadcast for autonomous driving. IEEE Wirel. Commun. **26**(4), 46–53 (2019)

22. M. Hadded, P. Muhlethaler, A. Laouiti, R. Zagrouba, L.A. Saidane, TDMA-based MAC protocols for vehicular ad hoc networks: a survey, qualitative analysis, and open research issues. IEEE Commun. Surv. Tut. **17**(4), 2461–2492 (2015)

23. R. Zhang, X. Cheng, L. Yang, X. Shen, B. Jiao, A novel centralized TDMA-based scheduling protocol for vehicular networks. IEEE Trans. Intel. Transport. Syst. **16**(1), 411–416 (2015)

24. A. Ahizoune, A. Hafid, R.B. Ali, A contention-free broadcast protocol for periodic safety messages in vehicular ad-hoc networks, in *IEEE Local Computer Network Conference* (2010), pp. 48–55

25. M.S. Almalag, S. Olariu, M.C. Weigle, TDMA cluster-based MAC for VANETs (TC-MAC), in *IEEE International Symposium on a World of Wireless, Mobile and Multimedia Networks (WoWMoM)* (2012), pp. 1–6

26. F. Borgonovo, A. Capone, M. Cesana, L. Fratta, ADHOC MAC: new MAC architecture for ad hoc networks providing efficient and reliable point-to-point and broadcast services. Wirel. Netw. **10**(4), 359–366 (2004)

27. W. Yang, P. Li, Y. Liu, H. Zhu, Adaptive TDMA slot assignment protocol for vehicular ad-hoc networks. J. China Univ. Posts Telecommun. **20**(1), 11–25 (2013)

28. X. Jiang, D.H.C. Du, PTMAC: a prediction-based TDMA MAC protocol for reducing packet collisions in VANET. IEEE Trans. Veh. Technol. **65**(11), 9209–9223 (2016)

29. W. Zhuang, Q. Ye, F. Lyu, N. Cheng, J. Ren, SDN/NFV-empowered future IoV with enhanced communication, computing, and caching. Proc. IEEE **108**(2), 274–291 (2020)

30. F. Lyu, H. Zhu, N. Cheng, H. Zhou, W. Xu, M. Li, X. Shen, Characterizing Urban vehicle-to-vehicle communications for reliable safety applications. IEEE Trans. Intell. Transp. Syst. **21**, 1–17. https://doi.org/10.1109/TITS.2019.2920813. Early Access, Jun. 2019

31. F. Bai, D.D. Stancil, H. Krishnan, Toward understanding characteristics of dedicated short range communications (DSRC) from a perspective of vehicular network engineers, in *Proceedings of the Sixteenth Annual International Conference on Mobile Computing and Networking* (2010)
32. F. Martelli, M. Elena Renda, G. Resta, P. Santi, A measurement-based study of beaconing performance in IEEE 802.11p vehicular networks, in *2012 Proceedings IEEE INFOCOM* (2012)
33. F. Bai, H. Krishnan, Reliability analysis of DSRC wireless communication for vehicle safety applications, in *Proceedings of IEEE Intelligent Transportation Systems Conference* (2006)
34. J. Gozalvez, M. Sepulcre, R. Bauza, IEEE 802.11p vehicle to infrastructure communications in Urban environments. IEEE Commun. Mag. **50**(5), 176–183 (2012)
35. I. Tan, W. Tang, K. Laberteaux, A. Bahai, Measurement and analysis of wireless channel impairments in DSRC vehicular communications, in *2008 IEEE International Conference on Communications* (2008)
36. L. Cheng, B. Henty, D. Stancil, F. Bai, P. Mudalige, Mobile vehicle-to-vehicle narrow-band channel measurement and characterization of the 5.9 GHz dedicated short range communication (DSRC) frequency band. IEEE J. Sel. Areas Commun. **25**(8), 1501–1516 (2007)
37. R. Meireles, M. Boban, P. Steenkiste, O. Tonguz, J. Barros, Experimental study on the impact of vehicular obstructions in VANETs," in *2010 IEEE Vehicular Networking Conference* (2010)
38. M. Boban, J. Barros, O. Tonguz, Geometry-based vehicle-to-vehicle channel modeling for large-scale simulation. IEEE Trans. Veh. Technol. **63**(9), 4146–4164 (2014)
39. N. Wisitpongphan, O.K. Tonguz, J.S. Parikh, P. Mudalige, F. Bai, V. Sadekar, Broadcast storm mitigation techniques in vehicular ad hoc networks. IEEE Wireless Commun. **14**(6), 84–94 (2007)
40. L. Briesemeister, G. Hommel, Role-based multicast in highly mobile but sparsely connected ad hoc networks, in *2000 First Annual Workshop on Mobile and Ad Hoc Networking and Computing. MobiHOC* (2000), pp. 45–50
41. Y.T. Yang, L.D. Chou, Position-based adaptive broadcast for inter-vehicle communications, in *IEEE International Conference on Communications Workshops* (2008), pp. 410–414
42. M. Li, W. Lou, K. Zeng, OppCast: Opportunistic broadcast of warning messages in VANETs with unreliable links, in *IEEE International Conference on Mobile Adhoc and Sensor Systems* (2009), pp. 534–543
43. W. Viriyasitavat, O.K. Tonguz, F. Bai, UV-CAST: an Urban vehicular broadcast protocol. IEEE Commun. Mag. **49**(11), 116–124 (2011)
44. H. Yoo, D. Kim, ROFF: RObust and fast forwarding in vehicular ad-hoc networks. IEEE Trans. Mobile Comput. **14**(7), 1490–1502 (2015)
45. F.J. Ros, P.M. Ruiz, I. Stojmenovic, Acknowledgment-based broadcast protocol for reliable and efficient data dissemination in vehicular ad hoc networks. IEEE Trans. Mobile Comput. **11**(1), 33–46 (2012)
46. O. Rehman, M. Ould-Khaoua, H. Bourdoucen, An adaptive relay nodes selection scheme for multi-hop broadcast in VANETs. Comput. Commun. **87**, 76–90 (2016)
47. M. Boban, R. Meireles, J. Barros, P. Steenkiste, O.K. Tonguz, TVR-Tall vehicle relaying in vehicular networks. IEEE Trans. Mobile Comput. **13**(5), 1118–1131 (2014)
48. H. Zhou, N. Cheng, Q. Yu, X. Shen, D. Shan, F. Bai, Toward multi-radio vehicular data piping for dynamic DSRC/TVWS spectrum sharing. IEEE J. Sel. Areas Commun. **34**(10), 2575–2588 (2016)
49. F. Lyu, N. Cheng, H. Zhu, H. Zhou, W. Xu, M. Li, X. Shen, Towards rear-end collision avoidance: adaptive beaconing for connected vehicles. IEEE Trans. Intel. Transport. Syst. 1–16. https://doi.org/10.1109/TITS.2020.2966586. Early Access, Jan. 2020
50. H. Zhou, N. Cheng, N. Lu, L. Gui, D. Zhang, Q. Yu, F. Bai, X. Shen, WhiteFi infostation: engineering vehicular media streaming with geolocation database. IEEE J. Sel. Areas Commun. **34**(8), 2260–2274 (2016)
51. H. Peng, X. Shen, Deep reinforcement learning based resource management for multi-access edge computing in vehicular networks. IEEE Trans. Netw. Sci. Eng. 1 (2020). https://doi.org/10.1109/TNSE.2020.2978856

52. M. Torrent-Moreno, J. Mittag, P. Santi, H. Hartenstein, Vehicle-to-vehicle communication: fair transmit power control for safety-critical information. IEEE Trans. Veh. Technol. **58**(7), 3684–3703 (2009)

53. J. Mittag, F. Schmidt-Eisenlohr, M. Killat, J. Härri, H. Hartenstein, Analysis and design of effective and low-overhead transmission power control for VANETs, in *VANET '08: Proceedings of the Fifth ACM International Workshop on VehiculAr Inter-NETworking* (2008), pp. 39–48

54. S.A.A. Shah, E. Ahmed, J.J.P.C. Rodrigues, I. Ali, R.M. Noor, Shapely value perspective on adapting transmit power for periodic vehicular communications. IEEE Trans. Intell. Transp. Syst. **19**(3), 977–986 (2018)

55. E. Egea-Lopez, P. Pavon-Mariuo, Fair congestion control in vehicular networks with beaconing rate adaptation at multiple transmit powers. IEEE Trans. Veh. Technol. **65**(6), 3888–3903 (2016)

56. C.L. Huang, Y.P. Fallah, R. Sengupta, H. Krishnan, Adaptive intervehicle communication control for cooperative safety systems. IEEE Netw. **24**(1), 6–13 (2010)

57. M. Sepulcre, J. Mittag, P. Santi, H. Hartenstein, J. Gozalvez, Congestion and awareness control in cooperative vehicular systems. Proc. IEEE **99**(7), 1260–1279 (2011)

58. E. Egea-Lopez, J.J. Alcaraz, J. Vales-Alonso, A. Festag, J. Garcia-Haro, Statistical beaconing congestion control for vehicular networks. IEEE Trans. Veh. Technol. **62**(9), 4162–4181 (2013)

59. A. Autolitano, C. Campolo, A. Molinaro, R.M. Scopigno, A. Vesco, An insight into decentralized congestion control techniques for VANETs from ETSI TS 102 687 V1.1.1, in *2013 IFIP Wireless Days (WD)* (2013), pp. 1–6

60. G. Bansal, J.B. Kenney, C.E. Rohrs, LIMERIC: a linear adaptive message rate algorithm for DSRC congestion control. IEEE Trans. Veh. Technol. **62**(9), 4182–4197 (2013)

61. T. Tielert, D. Jiang, Q. Chen, L. Delgrossi, H. Hartenstein, Design methodology and evaluation of rate adaptation based congestion control for vehicle safety communications, in *2011 IEEE Vehicular Networking Conference (VNC)* (2011), pp. 116–123

62. E. Egea-Lopez, P. Pavon-Mariuo, Distributed and fair beaconing rate adaptation for congestion control in vehicular networks. IEEE Trans. Mobile Comput. **15**(12), 3028–3041 (2016)

63. L. Zhang, S. Valaee, Congestion control for vehicular networks with safety-awareness. IEEE/ACM Trans. Netw. **24**(6), 3290–3299 (2016)

64. F. Lyu, N. Cheng, H. Zhou, W. Xu, W. Shi, J. Chen, M. Li, DBCC: leveraging link perception for distributed beacon congestion control in VANETs. IEEE Int. Things J. **5**(6), 4237–4249 (2018)

65. L. Sun, A. Huang, H. Shan, L. Cai, Adaptive beaconing for collision avoidance and tracking accuracy in vehicular networks, in *2017 IEEE Wireless Communications and Networking Conference (WCNC)* (2017), pp. 1–6

Chapter 3
Mobility-Aware and Collision-Avoidance MAC Design

In VANETs, the TDMA-based MAC protocol has been demonstrated as a promising solution to well support delay-sensitive and road-safety applications, since the time-slotted channel management can guarantee a medium access with determined delay. However, due to the varying vehicular mobilities, existing TDMA-based MAC protocols can result in collisions of time slot assignment when multiple sets of vehicles merge together. To avoid slot-assignment collisions, in this chapter, with considering the vehicular mobilities, we propose a mobility-aware TDMA-based MAC, named *MoMAC*, which can assign every vehicle a time slot according to the underlying road topology and lane distribution on roads. In *MoMAC*, different lanes on the same road segment and different road segments at intersections are associated with disjoint time slot sets. In addition, each vehicle broadcasts beacon application data together with the time slot occupying information of neighboring vehicles. By updating the time slot occupying information of two-hop neighbors (obtained indirectly from one-hop neighbors), vehicles can detect time slot collisions and access a vacant time slot in a fully distributed way. We demonstrate the efficiency of *MoMAC* with both theoretical analysis and extensive simulations. Compared with state-of-the-art TDMA-based MACs, the transmission collisions can be reduced by 59.2% when adopting our proposed *MoMAC*, and the rate of safety message transmissions/receptions can be significantly enhanced.

3.1 Problem Statement

Most road-safety applications rely on broadcast communications [1–3], and thus we need to carefully design a MAC protocol to support reliable one-hop broadcast, i.e., guaranteeing medium access delay with transmission collision avoidance. In the literature, various MAC protocols have been proposed for broadcast communication in VANETs, which can be categorized into the contention-based and contention-

© Springer Nature Switzerland AG 2020

F. Lyu et al., *Vehicular Networking for Road Safety*, Wireless Networks, https://doi.org/10.1007/978-3-030-51229-3_3

free MACs [4–7]. In general, contention-based MACs such as IEEE 802.11p are efficient when the number of contending vehicles is small. However, the access delay can increase to a significant level when the number of users becomes large with involving back off procedures. Therefore, the efficiency of the MAC protocols degrades significantly, especially in dense traffic conditions [8, 9]. In addition, for rapid response, the RTS/CTS scheme is disabled in broadcast mode, which can further aggravate the hidden terminal problem. Intuitively, it is challenging to design an efficient MAC protocol for reliable broadcast services under realistic VANETs. First, to enable high-priority road-safety applications, road-safety beacons need to be periodically exchanged with a high frequency, i.e., normally 10 Hz (every 100 ms) [10], which poses great pressures on medium resource management to guarantee medium access delay. Second, due to the variable network topology, diverse spatial densities of vehicles, and the hidden/exposed node problems, the MAC protocol has to be robust whenever and wherever, i.e., seamlessly adapting to dynamic communicating environments [11, 12]. Third, the lack of infrastructures in VANETs makes it hard to obtain the global network information and achieve fine-grained coordinations.

To cope with the above limitations, TDMA-based MAC protocols have been proposed in VANETs [4–6], since a predefined time slot usage is suitable for periodical broadcasting in a distributed manner. In TDMA-based MACs, time is partitioned into frames, each of which has a constant number of equal-length time slots. Time slots are synchronized among vehicles, and each vehicle is granted to access the channel at least once in each frame by occupying a distinct time slot. In existing TDMA-based MACs, if each vehicle is assigned with a unique time slot ideally, the stringent delay requirement of delivering road-safety beacons can be well guaranteed since no transmission collision would happen. However, in real-world driving scenarios, as vehicles move constantly, the performance of those protocols can deteriorate significantly due to the collisions of slot assignment when multiple sets of vehicles merge together, which is called *merging collision*. Figure 3.1 shows how merging collisions occur due to the vehicle mobility when adopting existing TDMA-based MACs. Specifically, in Fig. 3.1a, vehicles are moving in lanes with different speed limits towards the same direction. At the initial

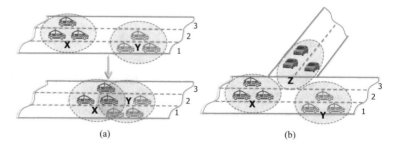

(a)   (b)

Fig. 3.1 Merging collisions caused by mobilities in prior TDMA-based MACs. (**a**) With diverse lane-speed limits. (**b**) At road intersections

stage, the vehicle set **X** and set **Y** are separate from each other, and vehicles in each set occupy a unique time slot for data transmission. As vehicles in set **X** move faster and catch up with vehicles in set **Y**, two sets then overlap, making it possible that vehicles in two sets use the same time slot, resulting in collisions. Figure 3.1b shows how merging collisions happen at an intersection. When a previously independent vehicle set **Z** approaches the intersection, it overlaps with the vehicle set **Y** and thus, merging collisions happen. To make things worse, if the vehicle set **Z** stops at the intersection due to the red light, it will continuously collide with all sets traversing in front of it. Another significant side effect of red traffic lights is that they make vehicles slow down until completely stop, which means that incoming vehicle sets on the blocked road join the merging sets at the intersection, leading to more severe merging collisions. Contradictorily, it is intersections that have the greatest need for reliable data communication to guarantee road safety. Therefore, it is essential to take into account those identified merging collisions when designing TDMA-based MAC protocols for VANETs.

Unlike other types of mobile users, the vehicular mobility is somewhat regulated, as the movement is constrained by both road layout and traffic rules, e.g., road signs, traffic lights, etc. It is potential to take advantage of such regulations of vehicular mobility to reduce the merging collisions [13–17]. We have the following two critical observations from the real-world vehicular environments: (1) vehicles may converge and diverge from time to time due to their distinct velocities and routes; (2) the design of the road topology and lane layout can statistically reflect the actual mobility demands of vehicles. Specifically, vehicles in the same lane pose relatively similar mobility patterns. If the vehicle wants to speed up or slow down, it will first choose to change a lane. In contrast, vehicles moving in the fast lane can always catch up with vehicles in the slow lane. Likewise, vehicles can eventually merge together at the intersection due to the road topological restrictions. Inspired by this, we propose *MoMAC*, an innovative mobility-aware TDMA-based MAC protocol, which can assign time slots elegantly according to the underlying road topology and lane distribution on roads. In *MoMAC*, different lanes on the same road segment and different road segments at intersections are associated with *disjoint time slot sets*, i.e., assigning vehicles that are bound to merge, with disjoint time slot sets. The merit of this design is that each vehicle can easily obtain the collision-avoidance slot assignment as long as the vehicle has a lane-level digital map and knows its current positional information of belonging to which road and which lane, which can be easily obtained by all the navigation systems [18–20]. To achieve a common agreement about the usage of slots among neighboring vehicles, we propose a fully *distributed* time slot assignment scheme, in which each vehicle selects a free time slot according to the received slot-occupying information from neighboring vehicles. Specifically, in addition to application data, each vehicle also broadcasts the information with IDs of its one-hop neighboring vehicles and slot indexes used by them. In doing so, transmission collisions can be detected by verifying the consistency of messages from neighbors, and all vehicles can keep pace with the link state changes in the moving. We analyze the performance of *MoMAC* theoretically in terms of average collisions and medium access delay. In addition, we conduct extensive simulations considering various road topologies and traffic conditions, and the results demonstrate the efficiency of MoMAC by checking the metrics of collision rate and safety message transmission/reception rate. The main contributions are summarized as follows.

- We identify two common mobility scenarios, which can result in massive merging collisions, while the existing TDMA-based MACs do not consider and cannot handle them well.
- We design a mobility-aware TDMA-based MAC, named *MoMAC*, to enhance the reliability of road-safety message exchange for road-safety applications. In *MoMAC*, the medium resource is assigned according to the underlying road topology and lane distribution on roads. By adopting our proposed *MoMAC*, merging collisions caused by vehicles' mobilities can be relieved.
- Both theoretical analysis and extensive implementation simulations are carried out to demonstrate the efficiency of *MoMAC*. In addition, the medium access delay and packet overhead are analyzed to verify the feasibility of *MoMAC*.

The remainder of this chapter is organized as follows. We present the system model in Sect. 3.2. Section 3.3 elaborates on *MoMAC* design. Performance analysis is carried out in Sect. 3.4. We conduct extensive simulations to evaluate the performance of *MoMAC* in Sect. 3.5. Section 3.6 gives a brief summary.

3.2   System Model

As shown in Fig. 3.2, we consider a system model including three main parts, i.e., wireless communication unit, road geography unit, and vehicles.

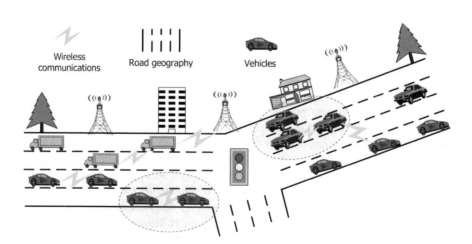

Fig. 3.2 Illustration of the system model

Wireless Communication All entities in the network communicate via Dedicated Short Range Communications (DSRC), which contains one Control Channel (CCH) and multiple Service Channels (SCHs) with two optional bandwidths of 10 and 20 MHz [21–23]. The CCH is essential and used to transmit high-priority short messages (such as periodic or event-driven road-safety messages) and control information (such as negotiation of SCHs usage among vehicles), while SCHs are used for user applications. In this chapter, we concentrate on the design of an efficient and reliable MAC protocol running on the CCH, which is the cornerstone for road-safety applications and multi-channel operations. To control the medium access on the CCH, time is partitioned into frames, each consisting of a given number S of fixed duration time slots, and each second contains an integer number of frames. To access the medium, a node has to be granted with a vacant time slot in the frame before it can transmit messages. In addition, the channel is considered to be symmetric, which has been evaluated by analyzing collected real-world DSRC communication data [24]. Thus, a node x is in the communication range of node y if and only if node y is in the communication range of node x.

Road Geography We consider real-world road scenarios involving highways and urban surface roads. We refer to a *road segment* as the road segment in one direction partitioned by two adjacent intersections. Road segments can have multiple lanes with different speed limits and are interconnected by intersections with traffic lights. We allow vehicles to have distinct acceleration and deceleration performance and to take actions such as overtaking or changing lanes whenever necessary.

Vehicles Vehicles in the network have at least one DSRC radio operating on the CCH, and they have identical communication capability and the same communication range R. For road safety, each vehicle has to broadcast its status information every 100 ms according to the requirement of road-safety applications [10]. Each vehicle is equipped with a GPS receiver that provides time reference and location information. For specific, the 1 pulse per second (PPS) signal provided by GPS receivers is used as a global time reference to synchronize vehicles. The rising edge of this 1 PPS is aligned with the start of every GPS second with accuracy within 100 ns even for low-end GPS receivers [4]. Hence, at any instant, each node can determine the index of the current slot within a frame. In addition, each vehicle has a lane-level digital map of the area of interest. Through GPS, each vehicle can obtain the positional information of which road and which lane it belongs to, matured in all the navigation systems [18–20]. Unlike previous position-guided MAC protocols which rely on the precise location information of vehicles, *MoMAC* adopts the underlying road topology and lane distribution on roads into medium resource allocation. It is practical as the road topology and lane distribution are constant and easily obtained. Furthermore, a vehicle just needs to know which road and which lane it belongs to, which are classification problems rather than an absolute positioning problem. As classifying a GPS location to a specific lane and road segment can tolerate localization errors ranging from meters to hundreds of meters, the inaccurate localization and temporary GPS shortage have a slight impact on the performance of the scheme.[1]

To facilitate time slot assignment in a distributed way, a vehicle x needs to maintain the information of its neighboring vehicles in one-hop and two-hop ranges, and the information lists are as follows:

- $N_{cch}(x)$: the set of IDs of its one-hop neighbors, which are updated by whether the node x has received packets directly on the channel during the previous S slots. In addition, the node x needs to broadcast this information with application data during each transmission.
- $N_{cch}^2(x)$: the set of IDs of its two-hop neighbors, indirectly obtained from the packets transmitted by its one-hop neighbors, i.e.,

$$N_{cch}^2(x) = N_{cch}(x) \cup \{N_{cch}(y), \forall y \in N_{cch}(x)\}.$$

- $U(x)$: the set of time slots that have been used by vehicles in the set of $N_{cch}^2(x)$.
- $G(x)$: the set of time slots pre-defined by *MoMAC* according to the current position of x, which is the possible set of time slots that x can choose from. $G(x)$ would be updated when x changes its mobility such as changing a lane, approaching or leaving an intersection (elaborated in the next section).
- $A(x)$: the available set of time slots that x currently can choose to use in the next frame. It is obtained based on the sets $U(x)$ and $G(x)$, i.e., $A(x) = G(x) - U(x)$.

3.3 MoMAC Design

3.3.1 Preliminaries About TDMA-Based MACs

In TDMA-based MAC protocols, time is partitioned into frames, each containing a fixed number of time slots. When using the TDMA-based MAC protocol, vehicles are synchronized via the GPS, and every vehicle is assigned with a time slot in each frame before it can transmit messages. Once a vehicle obtains a time slot successfully, it can use the same slot in all subsequent frames until a transmission collision is detected. In such protocols, neighboring vehicles within the communication range of the vehicle constitute its *one-hop set* (OHS). If two OHSs overlap with each other, the union of this two OHSs is referred to as a *two-hop set* (THS), in which each node can reach any other nodes in two hops at most. Figure 3.3 illustrates an example where the respective OHSs of vehicle A and vehicle C form a THS with vehicle B standing in both OHSs.

---

[1]In addition, to support future autonomous driving, the usage of HD (high definition) map is necessary, which can easily provide lane-level position information.

Fig. 3.3 Illustration of the
THS and hidden terminal
problem

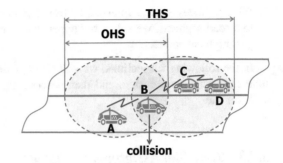

Obviously, vehicles in the same OHS should select different time slots to transmit messages. Moreover, vehicles in the same THS should also choose distinct time slots for communications in order to overcome the hidden terminal problem. The hidden terminal problem can arise in a THS when two vehicles, locating in two respective OHSs, cannot hear each other and decide to transmit a message in parallel. Taking an example in Fig. 3.3, vehicle A wants to transmit a message to vehicle B, and vehicle C wants to transmit a message to vehicle D at the same time. As vehicle A is not within the communication range of vehicle C, vehicle C would perceive that the channel is free, and start to transmit even though vehicle A has already started the transmission. As a result, there is a collision at vehicle B. To eliminate the hidden terminal problem when there is no RTS/CTS mechanism, each vehicle should collect (passively hear) and broadcast time slot-occupying information of one-hop neighbors, so that vehicles in one THS can know all the occupied time slots and detect possible collisions. As in the above example, since vehicles A and C transmitted simultaneously and caused the collision at the vehicle B, the IDS of A and C will not be included in $N_{cch}(B)$. if vehicle B broadcasts this information together with the application data, vehicle A (and C) can detect the collision since $B \in N_{cch}(A)$ but $A \notin N_{cch}(B)$.

3.3.2 Design Overview

In driving scenarios, vehicles may converge and diverge from time to time due to different velocities and routes. The collisions of time slot assignment can arise when vehicles merge in the moving. Therefore, we need to assign disjoint time slots to those vehicles that are bound to merge, which is the key operation of *MoMAC*. Specifically, the time slot assignment in *MoMAC* has the following three focuses:

1. When vehicles move on a multi-lane road segment, we leverage the lane distribution information on road to divide time slot sets.
2. When vehicles are at an intersection, we utilize the topology of the intersection, which converged by directional road segments, to divide time slot sets.

3. When vehicles enter an intersection from a road segment or leave an intersection to a road segment, we splice the upper two schemes together according to the geographical connection.

In the following subsections, we will first elaborate on the three focuses in the time slot assignment design, and then describe the time slot access approach for each vehicle.

3.3.3 Time Slot Assignment Scheme

We divide complex road network into individual road segments and intersections, and assign time slots for each of them so as to minimize potential collisions caused by vehicular mobilities.

On Multi-Lane Road Segments We partition each frame into three sets of time slots, i.e., L, R, and F as shown in Fig. 3.4. The F set is associated with RSUs, while the L and R sets are associated with road segments in left and right directions, respectively. As shown in Fig. 3.4, a road segment is said to be a left (right) road segment if it heads to any direction from north/south to west (east).

As vehicles moving in different lanes in the same direction can also cause merging collisions, in *MoMAC*, sets L and R are further divided into l subsets according the number of lanes l in that direction, i.e., $L_1, L_2, \ldots, L_l$ and $R_1, R_2, \ldots, R_l$. The subset L_i and R_i, $i \in [1, l]$ is assigned to the ith lane in left and right direction respectively, counted from the right direction. In practical, the l can be different in left and right directions. For example, as shown in Fig. 3.4, set R on a 3-lane road segment in the right direction is further split into three subsets, i.e., R_1, R_2, and R_3, each adopting one disjoint time slot subset. Notice that, in practical, due to the technical issue, the system sometimes may obtain the inaccurate lane

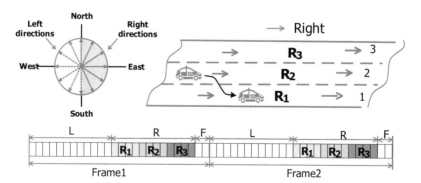

Fig. 3.4 Frames are divided into three slot sets, i.e., L, R, and F; set R on a 3-lane road in the right direction is further split into three subsets, i.e., R_1, R_2, and R_3, each using one disjoint slot set

information or miss the lane-changing detection. However, it has slight effect on *MoMAC*, as each node holds the frame information of its THS, which can assist the node in choosing a free time slot. Moreover, *MoMAC* can still work like VeMAC in the worst case, when without lane-level positional information.

At Intersections To eliminate merging collisions happened at intersections shown in Fig. 3.1b, we assign a separate time slot set for each road segment entering the intersection, called an *inbound* road segment. More specifically, given a *n*-way intersection, a frame is partitioned into $n + 1$ disjoint sets of time slots, i.e., ①, ②, ..., ⑨, and F. The F set is associated with RSUs, while set ⑧, $k \in [1, n]$, is assigned to the kth road segment entering the intersection counted anticlockwise from the north direction. In addition, set ⑧, $k \in [1, n]$, is also assigned to the kth road segment leaving the intersection, called *outbound* road segment, counted anticlockwise from the south direction. Figure 3.5 illustrates the slot assignment schemes for different types of intersections. For example, a three-way intersection is shown in Fig. 3.5a. As the road segment in the north-south direction enters the intersection (denoted by a solid arrowed line) and is the first road segment counted anticlockwise from the north, set ① is associated with this road segment. In contrast, as the road segment in the opposite direction (i.e., in the south-north direction) leaves the intersection (denoted by a dashed arrowed line) and is the second road segment counted anticlockwise from the south, set ② is associated to

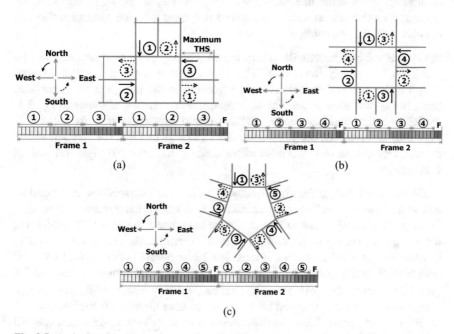

Fig. 3.5 Assigning time slots at intersections. (**a**) Three-way intersection. (**b**) Four-way intersection. (**c**) Five-way intersection

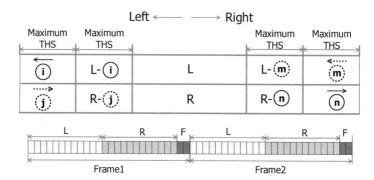

Fig. 3.6 Splicing road segments with intersections

the road segment. The same assign scheme applies to other types of intersections such as four-way intersections and five-way intersections, as shown in Fig. 3.5b and c, respectively.

Furthermore, as the density of vehicles at intersections can become heavy because of traffic control such as traffic lights and speed limits, we do not further divide set (k), $k \in [1, n]$, according to lanes at intersections in order to fully utilize time slots in set (k). To guarantee that any two neighboring road segments connecting to the same intersection are collision-free, as shown in Fig. 3.5a, the range of time slot sets on each road segment is defined to be the maximum size of a possible THS, i.e., equaling $2R$.

Splicing Road Segments with Intersections For any given road segment and the corresponding intersections associated with the road segment, it is straightforward to splice the slot assignment schemes according to the geographical connection between the road segment and the intersections. For example, as shown in Fig. 3.6, the left-direction road segment is partitioned into three parts and associated with three time slot sets, i.e., set (i) associated with a range of $2R$ at the very left end of the road segment, set (m) associated with a range of $2R$ at the very right end, and set L in between.

On one hand, due to the high density of vehicles at intersections, it is possible that vehicles at intersections can contend for time slots if the number of time slots in set (k), $k \in [1, n]$, is not sufficient. On the other hand, the density of vehicles at the middle of road segments tends to be low. To mitigate the slot shortage issue at intersections, in *MoMAC*, two extra ranges of $2R$ are added. For example, for the left road segment in Fig. 3.6, if set L and set (m) have common subsets, a range of $2R$ associated with set $L - (m)$ is arranged after set (m). The purpose of this arrangement is to release time slots occupied by vehicles that have already left the intersection, so that there are more free time slots available in set (m) for vehicles that are still at the intersection. Similarly, a range of $2R$ associated with set $L -$ ⓘ is arranged before set ⓘ.

3.3.4 Time Slot Access Approach

In this subsection, we describe how vehicles access time slots in our proposed *MoMAC*. During the transmission on the control channel, in the header of each packet, the transmitting node x should include set $N_{cch}(x)$ and the time slot used by each node $y \in N_{cch}(x)$. When a node x needs to acquire a time slot, it firstly listens to the channel for S consecutive time slots (not necessarily in the same frame). At the end of the S slots, the node x can obtain the information of $N_{cch}^2(x)$ and $U(x)$. As the set of $G(x)$ can be achieved based on the information of which road and which lane the vehicle belongs to, the node x can derive set $A(x) = G(x) - U(x)$ and randomly choose a time slot t from set $A(x)$ to use. After the node x transmits at the time slot t, it listens to the next $S - 1$ slots to determine whether the attempt to acquire the time slot t is successful. If packets received from all $z \in N_{cch}(x)$ indicate that $x \in N_{cch}(z)$, it means that there is no other node in the two-hop ranges of x attempting to access the same slot t. Under this condition, node x has successfully acquired the time slot t, and each node $z \in N_{cch}(x)$ adds x to its $N_{cch}(z)$ and updates the corresponding set $U(z)$. Otherwise, there is at least one node within the two-hop range of node x contending for the time slot t, and collisions (called *access collisions*) happen. As a result, all nodes contending for the time slot t are failed, and each has to acquire a new time slot until succeeds. Likewise, at the end of each time slot, each node x can actively perform the collision detection by checking a received packet from node $y \in N_{cch}(x)$. If the packet indicates that $x \notin N_{cch}(y)$, it means that the transmission from node x collides at node y with other concurrent transmissions. Once a collision is detected, node x has to release its time slot and try to apply for a new time slot.

In addition, in *MoMAC*, a node x needs to actively release its time slot and acquire a new one whenever necessary, in order to adapt to its real-time driving conditions, such as changing a lane and entering/leaving an intersection. As shown in Fig. 3.4, assume that a vehicle x moves in the second lane of a right-direction road segment, and thus the current $G(x)$ is R_2. Later, when vehicle x changes the lane from the second lane to the first lane (e.g., about to make a right turn), its $G(x)$ now changes to R_1. Based on the latest updated $U(x)$, if there is a free time slot in R_1, it releases the original time slot in R_2 and chooses a new time slot in R_1 for transmission. Otherwise, it would keep its time slot in R_2 until a free time slot in R_1 is available, or a collision with another vehicle in the second lane using the same time slot is detected. In doing so, merging collisions caused by vehicular mobilities can be significantly reduced. In *MoMAC*, although vehicles have to change slot frequently, it would not cause extra overhead cost to the original TDMA schemes. In addition, all decisions can be made based on the information collected from one-hop neighbors and the local positional information.

3.4 Performance Analysis

In this section, we conduct performance analysis of *MoMAC* by theoretically evaluating the average number of collisions with adopting *MoMAC* as well as other existing TDMA-based MACs. Besides, the medium access delay and communication overhead in *MoMAC* are modeled and analyzed to further verify its feasibility.

3.4.1 Average Number of Collisions

Existing TDMA-based MAC protocols focus on how to acquire time slots, detect collisions, and reapply for time slots after collisions. What *MoMAC* making difference is, to deal with the slot assignment collisions when vehicles merging due to diverse mobilities. In this subsection, we theoretically analyze the average number of collisions with existing TDMA-based MACs and *MoMAC*, respectively.

Collisions in Existing TDMA-Based MACs When adopting existing TDMA-based MACs, as shown in Fig. 3.7a, considering there are n ($n \geq 2$) vehicle sets, N time slots in each frame, and K vehicles in each set accessing time slots, vehicles in different sets are about to merge. In road-safety applications, the system has to assign each vehicle with a unique time slot, and thus we consider $N \geq nK$. Various vehicles from different vehicle sets may access the same time slots when the n vehicle sets merge at intersections or on multi-lanes roads, which could result in extensive merging collisions. Given a specific time slot in the frame, the probability of 0 and 1 vehicle accessing the time slot is denoted by p_0 and p_1, respectively. In existing TDMA-based MACs, they satisfy

$$p_0 = (\frac{C_{N-1}^K}{C_N^K})^n = (1 - \frac{K}{N})^n, \tag{3.1}$$

$$p_1 = C_n^1(\frac{C_{N-1}^{K-1}}{C_N^K})(1 - \frac{C_{N-1}^{K-1}}{C_N^K})^{n-1} = n\frac{K}{N}(1 - \frac{K}{N})^{n-1}. \tag{3.2}$$

Then, a collision would happen at the time slot if two or more vehicles are using the time slot simultaneously, the probability of which could be represented by $(1 - p_0 - p_1)$. We consider i ($i \leq \lfloor \frac{nK}{2} \rfloor$) time slots among the total N time slots are encountering the collisions, the probability of which is denoted by $P(i)$. The $P(i)$ satisfies

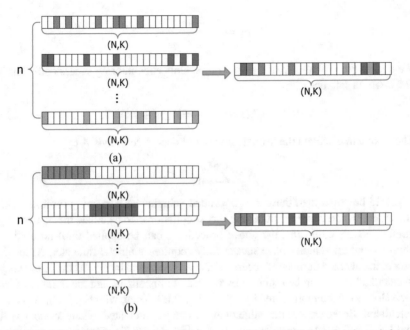

Fig. 3.7 Merging collisions happen, where there are n vehicle sets, N time slots in each frame, and K vehicles accessing time slots. In addition, white blocks denote free time slots while colorful blocks denote occupied time slots. (**a**) Merging collisions in existing MACs. (**b**) Merging collisions in *MoMAC*

$$P(i) = C_N^i (1 - p_0 - p_1)^i (p_0 + p_1)^{N-i}. \tag{3.3}$$

The average number of merging collisions is denoted by M, which satisfies

$$M = \sum_{i=1}^{\lfloor \frac{nK}{2} \rfloor} i * P(i). \tag{3.4}$$

Considering $2 * M$ vehicles have encountered merging collisions, these vehicles have to reapply for a new time slot, which could result in new access collisions.[2] To model the process, we can resort to the problem that there are $2 * x$ vehicles contending for y ($2x \le y$) free time slots. Given a specific time slot among y time slots, the probability of 0 and 1 vehicle using the time slot is denoted by p_0 and p_1, respectively, which satisfies

$$p_0 = \frac{(y-1)^{2x}}{y^{2x}}, \tag{3.5}$$

---

[2]Note that, to make the analysis tractable, we only consider the two-vehicle collision cases, since three- or more- vehicle collision cases rarely happen, and are rarely seen in our simulation results.

$$p_1 = \frac{(y-1)^{2x-1}}{y^{2x}}. \tag{3.6}$$

We denote the probability of i ($i \leq x$) time slots encountering access collisions by $P(i)$, which satisfies

$$P(i) = C_y^i (1 - p_0 - p_1)^i (p_0 + p_1)^{y-i}. \tag{3.7}$$

Then, we can calculate the average number of access collisions A by

$$A = \sum_{i=1}^x i * P(i). \tag{3.8}$$

It should be noted that, these $2 * A$ collided vehicles have to apply for a new time slot again, which could result in new access collisions. To calculate the new average number access collisions, the above procedures can be applied until no collision happens and all vehicles have successfully acquired a unique time slot. At the jth frame, the average number of access collisions is denoted by A_j. At the initial stage, to calculate A_1, x can be replaced by merging collisions M and the number of free time slots y can be represented by $(N - nK + 2M)$. Then, with Eq. (3.8), A_1 can be calculated. Based on A_j, the value of A_{j+1} can be calculated, where x and y is A_j and $(N - nK + 2A_j)$, respectively. To this end, denote T the total average number of collisions, which can be obtained via

$$T = M + \sum_{j=1} A_j. \tag{3.9}$$

Collisions in *MoMAC* As described above, in *MoMAC*, time slots are rescheduled in accordance with the vehicle mobilities and road topologies before vehicles merging together. As shown in Fig. 3.7b, vehicles are scheduled to adopt the disjoint time slot sets before the n vehicle sets merging together. With doing so, the system can eliminate the merging collisions M. However, vehicles have to release their time slots actively and apply for a new time slot when they change the mobility, such as leaving intersection. This process can result in additional access collisions. To this end, we consider the access collisions happened at an intersection where vehicles would leave the intersection. Likewise, for each road segment, the initial x and y can be represented by $\frac{(n-1)*K}{2*n}$ and $\frac{N-K}{n}$, respectively,[3] which can be used to calculate the value of A_1. Based on A_j, x and y can be represented by A_j and $(\frac{N-K}{n} - \frac{(n-1)*K}{n} + 2 * A_j)$, respectively, which can be used to calculate the value of A_{j+1}. Differently, for the total average number of access collisions, the value of n should be multiplied per the number of road segments. It can be calculated as

---

[3]Note that, to make the analysis tractable, we have a reasonable assumption that vehicles would enter each new road segment with an equal probability. Therefore, for each road segment, there are $\frac{(n-1)*K}{n}$ vehicles may enter and contend for the unoccupied $\frac{N-K}{n}$ time slots. It should be pointed out that, with the control of traffic lights, vehicles cannot enter the road segment simultaneously, and thus parameters are conservatively set here.

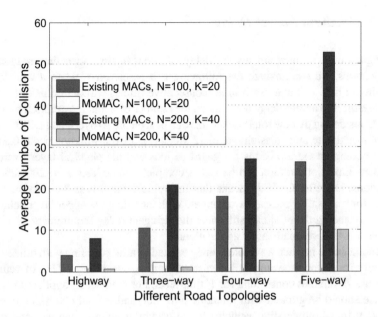

Fig. 3.8 Theoretical average number of collisions under different road topologies with different settings of N and K

$$T = n * \sum_{j=1} A_j.\tag{3.10}$$

Theoretical Results We plot the theoretical results of Eqs. (3.9) and (3.10) in Fig. 3.8, with varying the values of n, N and K. The following three major statements can be made. First, under all road topologies, *MoMAC* can achieve a better performance in terms of fewer collisions when with the same parameter setting. For example, when with n, N, and K being respective 5, 200, and 40, the average number of collisions is about 52.77 and 9.99 when adopting the existing MACs and MoMAC, respectively. It shows that more than 81.1% collisions can be reduced in *MoMAC*. Second, although in all MAC protocols, the average number of collisions increases with the road topology complexity, the increasing ratio is more significant in existing MACs and there is just a slight increase in *MoMAC*. It demonstrates that under different scenarios, *MoMAC* can work more robustly. Third, under all road topologies, the average number of collisions increases with N and K when adopting existing MACs, which however decreases with N and K when adopting *MoMAC*. It can be explained as that with larger values of N and K, much severe merging collisions can be triggered, which however can reduce the effect of access collisions. In addition, when adopting existing MACs, the merging collision has more negative effects than that of the access collision, which can be verified via Eq. (3.9).

3.4.2 Medium Access Delay

To figure out the medium access delay of *MoMAC* in supporting road-safety applications, we can analyze the following two cases. First, when a vehicle has acquired a time slot and use it in all consecutive frames without collisions, called *stable state*, the access delay depends on the number of S and the duration of a time slot. Considering that the total packet size of a beacon is 380 bytes, and DSRC radios adopt a moderate transmission rate of 12 Mbps [7], the transmission needs 0.25 ms. After adding an extra 0.05 ms for guard periods and the physical layer overhead, a 0.3 ms time slot duration can be set. A complete frame lasts $S = 200$ time slot durations, i.e., 60 ms, which means that the vehicle can access the medium every 60 ms for road-safety message exchanges. With including the upper layer delay and packet queueing delay, it can still meet the stringent delay requirement of 100 ms for most high-level road-safety applications.

Another case is when a vehicle (newly opened or after colliding with others) tries to apply for a new time slot, called *unstable state*. Considering a THS of vehicles, there are K vehicles contending for F time slots. For road-safety applications, each vehicle should be granted with a unique time slot, and we only consider $F \geq K$. In every frame, contending vehicles try to occupy a unique time slot and detect whether the trying acquisition is successful. If one vehicle successfully acquires a time slot in the frame, the vehicle will end the contending process and transfer to the stable state. Otherwise, the vehicle has to continue the contending process in the subsequent frame. Assuming that the number of contending vehicles K remains constant in the THS, the contending process can be modeled as follows. Let X_n be the number of vehicles that have successfully acquired a unique time slot at the end of the nth frame, and $X_0 = 0$ be the initial state. X_n can be then modeled as a stationary discrete-time Markov chain,[4] and the transition process is shown in Fig. 3.9 with the following transition probabilities,

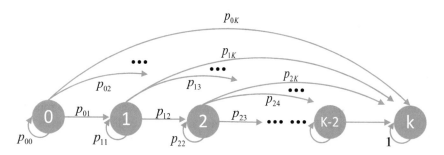

Fig. 3.9 The transition process of X_n

---

[4]Note that, the X_n has no possibility to be the value of $K - 1$, as one vehicle cannot collide with itself.

$$p_{ij} = \begin{cases} \frac{f(j-i,K-i,F-i)}{(F-i)^{K-i}} & \begin{aligned} & 0 \le i \le K-2, \\ & i \le j \le K; \end{aligned} \\[2ex] 1 & i = j = K; \\[2ex] 0 & \begin{aligned} & i > j \quad \text{or} \quad i = K-1 \\ & \text{or} \quad j = K-1, \end{aligned} \end{cases} \tag{3.11}$$

where $f(l, u, v)$ is the number of cases that for v available free time slots, l nodes successfully acquire a unique time slot among u contending nodes. To compute the value of $f(l, u, v)$, $l \le u \le v$, we can consider a scenario that there are u balls needed to be be packed into v boxes, and each box can support more than one balls. $f(l, u, v)$ can be the number of packing ways satisfying that existing l boxes only contain one ball, and the other $v - l$ boxes are either empty or contain more than one ball. Then the $f(l, u, v)$ satisfies

$$f(l, u, v) = \begin{cases} C_u^l A_v^l ((v-l)^{u-l} - \\ \sum_{i=1}^{u-l} f(i, u-l, v-l)) & 0 \le l < u; \\ A_v^l & l = u. \end{cases} \tag{3.12}$$

Based on this, the one-step transition probability matrix P can be computed. Let P^n be the n-step transition probability matrix, and the first row of P^n represent the distribution of X_n, i.e.,

$$p(X_n = i) = P_{1,i+1}^n, i \in [0, K]. \tag{3.13}$$

The probability that a specific vehicle successfully acquires a unique time slot within n frames is

$$p_{success} = \Sigma_{i=1}^K \frac{C_{K-1}^{i-1}}{C_K^i} p(X_n = i) = \frac{\Sigma_{i=1}^K i P_{1,i+1}^n}{K}. \tag{3.14}$$

Figure 3.10 shows the numerical results of Eq. (3.14). For better presentation, we introduce a coefficient a to describe the relationship between the number of contending nodes K and available free time slots F, i.e.,

$$F = aK. \tag{3.15}$$

We can have two major conclusions. First, when the resource is limited, collisions should be carefully avoided as they can incur severe access delays. For instance, when $K = 10$, to guarantee a more than 90% probability of vehicles successfully acquiring a unique time slot, 7 frames, 3 frames and 2 frames are required under the set of $a = 1$, $a = 2$ and $a = 3$, respectively. Second, under the same resource

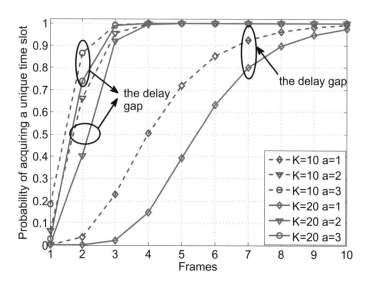

Fig. 3.10 The access delay under unstable state

condition, collisions can also affect the access delay significantly. Specifically, when $a = 1$, the probability can reach 72% at the 5th frame with the set of $K = 10$, but the probability can only be 39% at the 5th frame with the set of $K = 20$. As road-safety applications in VANETs have a quite low tolerance to the messages delivery delay, medium access delay as the dominating delay should be carefully guaranteed without collisions, which is also the main focus of *MoMAC*.

3.4.3   Packet Overhead

The main overhead of *MoMAC* is the needed coordination information for medium access, including the vehicle IDs and the corresponding time slot indexes of neighbors in the OHS. Let V_{max} and V^2_{max} be the maximum number of vehicles existing in an OHS and THS. The number of $\lceil \log_2 V^2_{max} \rceil$ bits are required to represent the individual ID of vehicles in the THS, where the symbol $\lceil . \rceil$ is ceil function. To identify a specific time slot among S slots, $\lceil \log_2 S \rceil$ bits are needed. Hence, the total overhead of *MoMAC* H (in bits) is

$$H = |N_{cch}(x)|(\lceil \log_2 V^2_{max} \rceil + \lceil \log_2 S \rceil). \tag{3.16}$$

As the maximum OHS area is a circle with the radius R, the V_{max} on a road can be computed by

$$V_{max} = (\frac{2R}{length_{vehicle} + distance_{safety}}) * L, \qquad (3.17)$$

where the $length_{vehicle}$ is the length of a vehicle, normally 3–5 m for sedans [25], $distance_{safety}$ is the safe following distances, and L is the number of lanes on the road. According to the 2 second rules, drivers should drive at least 2 s behind the front vehicle even with ideal conditions. Given a normal speed of 60 km/h in the urban environment, the $distance_{safety}$ can be obtained, i.e., $distance_{safety} \approx$ 35 m. Let $R = 300$ and $L = 6$ respectively for normal case, and then $|N_{cch}(x)| = V_{max} = (\frac{600}{5+35}) * 6 = 90$. To guarantee each vehicle with a unique time slot in a THS, we set the S and V_{max}^2 to be 200 empirically. The overhead in the case is $H = 90 * 16 = 1440$ bits ≈ 180 bytes. The size of application data broadcasted by road-safety applications is small, normally 200–500 bytes [26]. Adding such extra 180 bytes of coordination data in broadcast packets is *acceptable*, since the total packet size is far smaller than the size that the MAC protocol can support.

3.5 Performance Evaluation

In this section, to evaluate the performance of *MoMAC*, we conduct extensive simulations under various practical road topologies and traffic conditions.

3.5.1 Methodology

Simulation Setup To emulate the vehicular driving conditions, we adopt the Simulation of Urban Mobility (SUMO) [27] to create the transportation system, where two typical VANET environments are built, i.e., highway and urban road topologies. To be specific, for the highway environment, we build a 10 km long and bidirectional 8-lane highway scenario, where the speed limit for four lanes in one direction is respectively set to be 60 km/h, 80 km/h, 100 km/h, and 120 km/h. For the urban environment, we construct three different star topologies, which have different types of intersections, i.e., 3-way, 4-way, and 5-way, locating at the center, each respectively connecting to three, four, and five bidirectional 6-lane roads. In addition, each bidirectional road segment is 4 km long, and the speed limit for each three lanes in one direction is set to be 50 km/h, 60 km/h, and 70 km/h, respectively. Moreover, we set traffic lights at each inbound road segment of intersections, where the green light duration is set to be 20 s. It should be noted that, the star topologies are considered due to the reason of without losing generality, which can evaluate the system performance over different traffic conditions. Based on star topologies, more complicated topologies can be easily created.

Under both driving environments, we set vehicles with different running parameters, where the maximum velocity is ranged from 80 to 240 km/h, acceleration capability is ranged from 1 to 5 m/s², and the deceleration capability is ranged from 3 to 10 m/s², respectively. Ten different settings of vehicle parameters are configured per the main types of vehicles on the market, and each vehicle is randomly associated with one setting. We generate vehicles at the open end of each road segment with different rates to mimic different traffic conditions in a day, i.e., light traffic (3 vehicles/lane/min), moderate traffic (5 vehicles/lane/min), and heavy traffic (10 vehicles/lane/min). When the vehicle enters a road segment, it will randomly choose a moving lane and destination road segment. Besides, to conduct the speed and lane-change control, vehicles are driven under the Krauss car-following and LC2013 lane-changing model. In simulations, to mimic the human factor, the driver imperfection parameter is also integrated. The simulated snapshots of highways, three-way, four-way, and five-way intersections are shown in Fig. 3.11, where different parameter settings are represented by different vehicle colors.

For communication setting, we set the transmission range R to be 300 m in accordance with the observation that 802.11p-compatible onboard units can support reliable data transmission within 300 m [23]. In addition, all transmissions when within the communication range, are considered to be successful unless time slot usage collisions happen, as we focus on the MAC performance. To support road-safety applications, we set the frame duration to be 100 ms to comply with the rigid delay requirement of applications [10], and thus the number of time slots in each

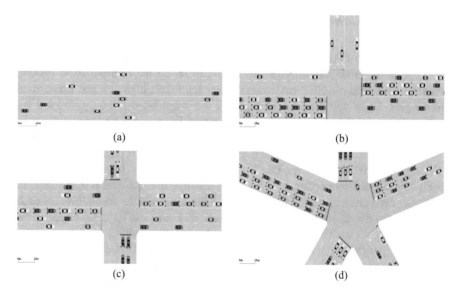

(a)

(b)

(c)

(d)

Fig. 3.11 Snapshots of the simulated scenarios. (**a**) Highway. (**b**) Three-way intersection. (**c**) Four-way intersection. (**d**) Five-way intersection

frame is set to be 200. For each round of simulation, we run experiments for 1500 s with logging all the system metrics.

Performance Metrics The following metrics are designed to evaluate the performance.

1. **Rate of collisions**: refers to the average number of transmission collisions per frame per THS.[5]
2. **Rate of safety message transmissions**: refers to the average number of successful beacon transmissions per frame per THS. When one vehicle broadcasts a beacon, the transmission is regarded to be successful if there is no concurrent transmissions within the THS.
3. **Rate of safety message receptions**: refers to the average number of successful beacon receptions per frame per THS. One beacon can be successfully received by the vehicle if there is no other beacon arrives at the vehicle simultaneously.

For performance comparison, we adopt the benchmarks of VeMAC [4] and ADHOC MAC [5].

3.5.2 Impact of Various Road Topologies

We first examine the impact of road topologies. The cumulative distribution functions (CDFs) of rate of collisions under different environments are shown in Fig. 3.12, where the moderate traffic condition is adopted. With the results, the following two major statements can be made. First, under all environments, *MoMAC* can achieve the best performance with the lowest collision rate. Second, when the road topology becoming complicated, the achieved performance under both the MAC of VeMAC and ADHOC can degrade, which, however, can remain stable under all environments when adopting our proposed *MoMAC*. For example, when under the highway, three-way intersection, four-way intersection, and five-way intersection, the probability that the transmission can be delivered without collision reaches about 85.9%, 82.8%, 76.8%, and 73.6% in *MoMAC*, respectively. However, the probability is reduced to 71.7%, 59.8%, 30.4%, and 28.3% in VeMAC when in respective driving environments, and the probabilities under all environments are less than 33.6% in ADHOC MAC. For complicated road topologies, more road segments can combine at the intersection, where more vehicle sets will merge and result in significant merging collisions. However, as there is only a slight increase of

---

[5]To get the metrics in per THS, the metric is first calculated for the whole simulation area, and then is multiplied by $\frac{2R}{L}$ or $\frac{2R}{L} \times \frac{1}{N}$ for the highway and urban environments, respectively, where $2R$ is the maximum THS length, L is the length of the road segment, and N is the number of inbound road segments at an intersection.

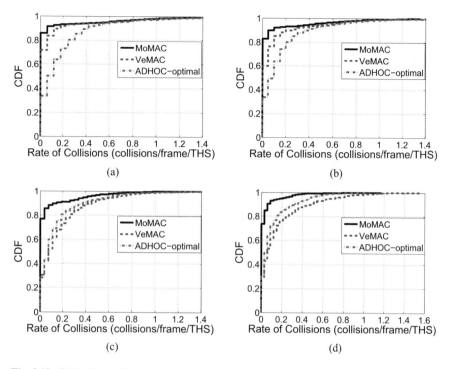

Fig. 3.12 CDFs of rate of collision in different environments under moderate traffic condition. (**a**) Highway. (**b**) Three-way intersection. (**c**) Four-way intersection. (**d**) Five-way intersection

transmission collisions in *MoMAC*, it can verify its robustness in adapting to diverse road topologies.

The CDFs of rate of safety message transmissions under different environments are shown in Fig. 3.13, where the moderate traffic condition is adopted. The following two major observations can be achieved. First, under all environments, *MoMAC* can achieve the best performance with the supreme rate of safety message transmissions. Second, the performance gap between *MoMAC* and other two benchmarks can increase with the road complexity. For instance, as shown in Fig. 3.13a, when under the highway environment, the CDF results of safety message transmission rates achieved by three protocols are tightly closed, but when under the three-way intersection, four-way intersection, and five-way intersection, the performance gap between *MoMAC* and other two benchmarks shows up, especially as shown Fig. 3.13d, the performance gap is rather obvious. It can be explained since the side effects of transmission collisions on VeMAC and ADHOC MAC are more serious than that of on *MoMAC* when the road topology becomes complicated and more vehicles are bound to merge. Figure 3.14 shows the CDFs of rates of safety message receptions under all environments with the moderate traffic condition, where we can observe the similar observations. Differently, as we investigate the broadcasting activities, *MoMAC* can achieve much more benefits in

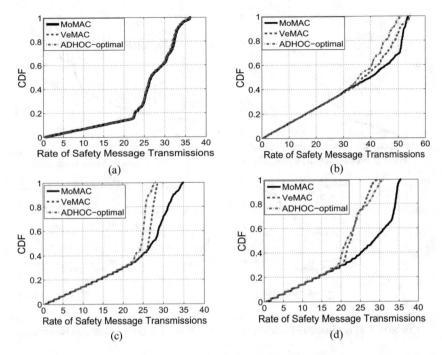

Fig. 3.13 CDFs of rate of safety message transmissions in different environments under moderate traffic condition. (**a**) Highway. (**b**) Three-way intersection. (**c**) Four-way intersection. (**d**) Five-way intersection

terms of safety message reception rates since one transmission activity can result in various reception activities. To be specific, as shown in Figs. 3.13d and 3.14d, when examining the medium performance (i.e., with the CDF value of 0.5), the performance gap between *MoMAC* and benchmarks in terms of safety message transmission rate is about 10/frame/THS, but the performance gap in terms of safety message reception rate can reach above 1200/frame/THS. It indicates that when adopting *MoMAC*, more than 1200 beacon receptions can be successfully guaranteed every 100 ms in a THS, which is significant for the service quality of road-safety applications.

3.5.3 Impact of Dynamic Traffic Conditions

The impact of traffic conditions is then investigated in this subsection. Under the four-way intersection environment, we plot the CDFs of rate of collisions with different traffic conditions in Fig. 3.15. It can be seen that under all traffic conditions, *MoMAC* can achieve the best performance with the minimum number of transmission collisions. Additionally, we can observe that, with more traffic, although the

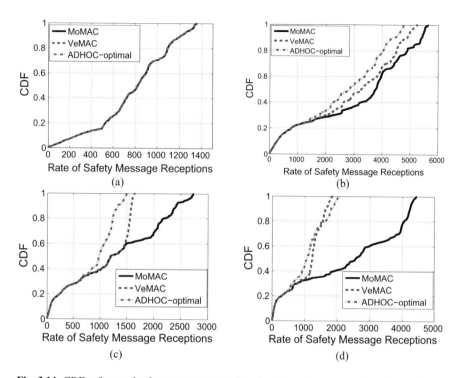

Fig. 3.14 CDFs of rate of safety message receptions in different environments under moderate traffic condition. (**a**) Highway. (**b**) Three-way intersection. (**c**) Four-way intersection. (**d**) Five-way intersection

achieved performance by all MACs degrades, the performance degrades slightly in *MoMAC*. For instance, when with the heaviest traffic condition shown in Fig. 3.15c, *MoMAC* can still work effectively where the collision-free probability reaches about 49.9%. However, the probability can reduce to 16.6% and 15.6% when adopting VeMAC and ADHOC MAC, respectively. The results demonstrate that VeMAC and ADHOC MAC generally have a poor performance when under heavy traffic conditions, but *MoMAC* can work robustly under all traffic conditions. Under the four-way intersection environment, we plot the CDFs of rate of safety message transmissions with different traffic conditions in Fig. 3.16, where we can achieve similar observations. To be specific, under all traffic conditions, *MoMAC* can achieve the best performance with the highest safety message transmission rates. In addition, the performance gap between *MoMAC* and the other two benchmarks becomes more significant with the traffic density. It demonstrates that compared with other two MACs, the heavy traffic condition has slighter side effects on *MoMAC*. We omit the results of safety message reception rates due to the similar observations.

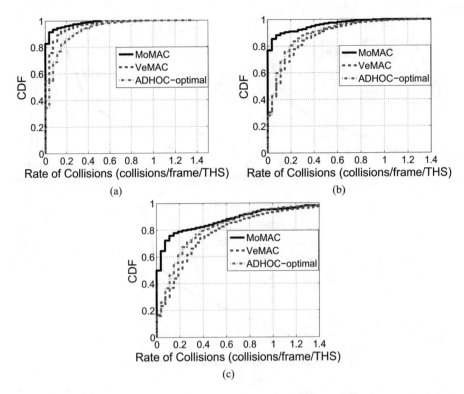

Fig. 3.15 CDFs of rate of collisions under different traffic conditions at the four-way intersection.
(**a**) Light traffic. (**b**) Moderate traffic. (**c**) Heavy traffic

3.6 Summary

In this chapter, to reduce transmission collisions in the moving, we have proposed
a mobility-aware TDMA-based MAC protocol for VANETs, named *MoMAC*. We
have first identified two common mobility scenarios that can result in massive transmission collisions in vehicular environments. A simple yet effective slot assignment
scheme then has been proposed, which can fully utilize the underlying road topology
and lane layout information, to reply to the potential communication demands per
vehicular mobilities. To eliminate the hidden terminal problem, *MoMAC* adopts a
fully distributed slot access and collision detection scheme. Theoretical analysis
and extensive simulation results have been carried out to demonstrate the efficiency
of *MoMAC*. Note that, the performance of *MoMAC* can be further enhanced by
leveraging RSUs to act as coordinators. They can calculate the current traffic
condition and make an optimal disjoint subsect division for uneven traffic, and
broadcast the assignment to vehicles in vicinity. *MoMAC* can fit in this solution

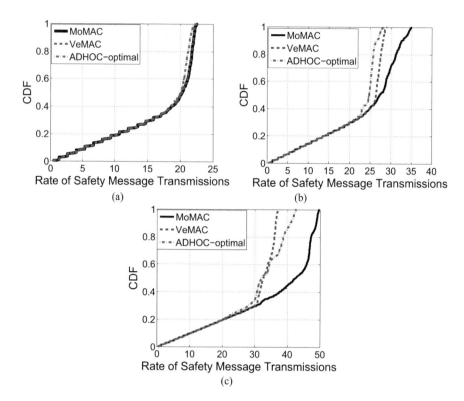

Fig. 3.16 CDFs of rate of safety message transmissions under different traffic conditions at the four-way intersection. (**a**) Light traffic. (**b**) Moderate traffic. (**c**) Heavy traffic

well as RSUs can listen to all broadcast messages and conduct statistics on the traffic condition on each road segment, and then broadcast out the up-to-date slot assignment.

References

1. F. Lyu, N. Cheng, H. Zhu, H. Zhou, W. Xu, M. Li, X. Shen, Towards rear-end collision avoidance: adaptive beaconing for connected vehicles. IEEE Trans. Intell. Transp. Syst. 1–16. https://doi.org/10.1109/TITS.2020.2966586. Early Access, Jan. 2020
2. W. Xu, H. Zhou, N. Cheng, F. Lyu, W. Shi, J. Chen, X. Shen, Internet of vehicles in big data era. IEEE/CAA J. Autom. Sin. **5**(1), 19–35 (2018)
3. N. Cheng, F. Lyu, J. Chen, W. Xu, H. Zhou, S. Zhang, X. Shen, Big data driven vehicular networks. IEEE Netw. **32**(6), 160–167 (2018)
4. L.L.H.A. Omar, W. Zhuang, VeMAC: a TDMA-based MAC protocol for reliable broadcast in VANETs. IEEE Trans. Mob. Comput. **12**(9), 1724–1736 (2013)

5. F. Borgonovo, A. Capone, M. Cesana, L. Fratta, ADHOC MAC: new MAC architecture for Ad Hoc networks providing efficient and reliable point-to-point and broadcast services. Wirel. Netw. **10**(4), 359–366 (2004)
6. F. Lyu, H. Zhu, H. Zhou, W. Xu, N. Zhang, M. Li, X. Shen, SS-MAC: a novel time slot-sharing MAC for safety messages broadcasting in VANETs. IEEE Trans. Veh. Technol. **67**(4), 3586–3597 (2018)
7. IEEE Std 802.11p-2010, Standard for Information Technology-Telecommunications and Information Exchange between Systems-Local and Metropolitan Area Networks-Specific Requirements Part 11: Wireless LAN Medium Access Control (MAC) and Physical Layer (PHY) Specifications Amendment 6: Wireless Access in Vehicular Environments, pp. 1–51, July 2010
8. W. Zhu, D. Gao, C.H. Foh, W. Zhao, H. Zhang, A collision avoidance mechanism for emergency message broadcast in urban VANET, in *IEEE 83rd Vehicular Technology Conference (VTC Spring)*, May 2016, pp. 1–5
9. H. Nguyen-Minh, A. Benslimane, D.-J. Deng, Reliable broadcasting using polling scheme based receiver for safety applications in vehicular networks. Veh. Commun. **4**, 1–14 (2016)
10. CAMP Vehicle Safety Communications Consortium and others, Vehicle Safety Communications Project: Task 3 Final Report: Identify Intelligent Vehicle Safety Applications Enabled by DSRC, in *National Highway Traffic Safety Administration, US Department of Transportation*, Washington, DC, March 2005
11. W. Zhuang, Q. Ye, F. Lyu, N. Cheng, J. Ren, SDN/NFV-empowered future IoV with enhanced communication, computing, and caching. Proc. IEEE **108**(2), 274–291 (2020)
12. Q. Ye, W. Zhuang, L. Li, P. Vigneron, Traffic-load-adaptive medium access control for fully connected mobile Ad Hoc networks. IEEE Trans. Veh. Technol. **65**(11), 9358–9371 (2016)
13. H. Zhou, N. Cheng, N. Lu, L. Gui, D. Zhang, Q. Yu, F. Bai, X. Shen, WhiteFi Infostation: engineering vehicular media streaming with geolocation database. IEEE J. Sel. Areas Commun. **34**(8), 2260–2274 (2016)
14. T. Taleb, E. Sakhaee, A. Jamalipour, K. Hashimoto, N. Kato, Y. Nemoto, A stable routing protocol to support ITS services in VANET networks. IEEE Trans. Veh. Technol. **56**(6), 3337–3347 (2007)
15. L. Yao, J. Wang, X. Wang, A. Chen, Y. Wang, V2X routing in a VANET based on the hidden Markov model. IEEE Trans. Intell. Transp. Syst. **19**(3), 889–899 (2018)
16. H. Zhou, N. Cheng, Q. Yu, X. Shen, D. Shan, F. Bai, Toward multi-radio vehicular data piping for dynamic DSRC/TVWS spectrum sharing. IEEE J. Sel. Areas Commun. **34**(10), 2575–2588 (2016)
17. X. Ge, J. Ye, Y. Yang, Q. Li, User mobility evaluation for 5G small cell networks based on individual mobility model. IEEE J. Sel. Areas Commun. **34**(3), 528–541 (2016)
18. Google maps for mobile. http://www.google.com/mobile/maps/
19. B. Wang, Q. Ren, Z. Deng, M. Fu, A self-calibration method for nonorthogonal angles between gimbals of rotational inertial navigation system. IEEE Trans. Ind. Electron. **62**(4), 2353–2362 (2015)
20. Z. Wu, J. Li, J. Yu, Y. Zhu, G. Xue, M. Li, L3: sensing driving conditions for vehicle lane-level localization on highways, in *Proceedings of IEEE INFOCOM*, July 2016
21. H. Zhou, W. Xu, J. Chen, W. Wang, Evolutionary V2X technologies toward the internet of vehicles: challenges and opportunities. Proc. IEEE **108**(2), 308–323 (2020)
22. Y. Li, An overview of the DSRC/WAVE technology, in *Quality, Reliability, Security and Robustness in Heterogeneous Networks* (Springer, New York, 2012), pp. 544–558
23. F. Lyu, H. Zhu, N. Cheng, H. Zhou, W. Xu, M. Li, X. Shen, Characterizing urban vehicle-to-vehicle communications for reliable safety applications. IEEE Trans. Intell. Transp. Syst. 1–17. https://doi.org/10.1109/TITS.2019.2920813. Early Access, June 2019
24. F. Bai, D.D. Stancil, H. Krishnan, Toward understanding characteristics of dedicated short range communications (DSRC) from a perspective of vehicular network engineers, in *Proceedings of ACM MobiCom*, September 2010

25. Wikipedia. Vehicle Size Class. Available: https://en.wikipedia.org/wiki/Vehicle_size_class
26. DSRC Committee, Dedicated Short Range Communications (DSRC) Message Set Dictionary. Society of Automotive Engineers, Warrendale, PA. Technical Report J2735_200911, November 2009
27. DLR Institute of Transportation Systems, SUMO: Simulation of Urban MObility. http://www.dlr.de/ts/en/desktopdefault.aspx/tabid-1213/

Chapter 4
Efficient and Scalable MAC Design

After resolving the mobility issue in medium access, we focus on the efficient and scalable MAC design in this chapter. Specifically, existing TDMA-based MACs do not consider the situation of diverse beaconing rates at vehicles, and such inflexible design may suffer from a scalability issue in terms of channel resource management. For instance, scarce channel resources can be wasted due to unnecessary broadcasting under light traffic densities, and the unfairness of medium resource allocation can aggravate message collisions under heavy traffic densities. In this chapter, we propose a novel time *Slot-Sharing MAC*, named *SS-MAC*, which can support diverse beaconing rates of vehicles. Particularly, we first introduce a circular recording queue to online perceive time slot occupying status. We then design a distributed time slot sharing (DTSS) approach and random index first fit (RIFF) algorithm, to efficiently share the time slot and conduct the online vehicle-slot matching, respectively. We prove the efficacy of DTSS algorithm theoretically and evaluate the efficiency of RIFF algorithm by using Matlab simulations. Finally, under various driving scenarios and resource conditions, we conduct extensive implementation simulations to demonstrate the efficiency of *SS-MAC* in terms of delay ratios of the overall system and individual vehicle.

4.1 Problem Statement

To enable reliable broadcast communications for road-safety applications, TDMA-based MACs have been demonstrated with efficiency in VANETs [1–5]. In TDMA-based MACs, time is partitioned into frames consisting of a constant number of equal-length time slots and synchronized among vehicle nodes. Each vehicle is guaranteed to access the channel at least once in each frame, and the slotted channel can guarantee the stringent time requirement of road-safety applications. However, in existing TDMA-based MACs, a high and fixed beaconing rate, normally 10 Hz, is

© Springer Nature Switzerland AG 2020
F. Lyu et al., *Vehicular Networking for Road Safety*, Wireless Networks,
https://doi.org/10.1007/978-3-030-51229-3_4

configured for all nodes, which can lead to a scalability issue for resource allocation. Specifically, when the node density is low and the road condition is bright for driving, scarce channel resources [6–8] can be wasted due to the unnecessary broadcasting. According to the vehicle safety communications report of the U.S. Department of Transportation [9], there are distinct road-safety applications with broadcasting rates ranging from 1 to 10 Hz. Allocating excess resources to each vehicle for unnecessary broadcasting can not only waste channel resources but also increase the possibility to interfere others. On the other hand, due to the spatial reuse constraint of time slot,[1] when the vehicle density is high, such as at intersections, the *slot shortage problem* may occur. As a result, the unfairness of medium resource allocation can happen among vehicles. For instance, when some vehicles at the intersection have fully occupied time slots in every frame, subsequent entering vehicles may have no time slot to choose from, and then the time slot acquisition failures can last for a long time, resulting in a significant medium access delay. To make things worse, if vehicles with low-priority[2] road-safety requirements have successfully occupied time slots, while vehicles with high-priority road-safety requirements have no chance for transmission, this unfairness situation of medium resource allocation can significantly impair the road safety. Just like traffic management in real life, when there is a traffic jam, the vehicles with more importance, e.g., police cars or ambulances, have a higher priority of passing through. The medium access control should also have this kind of scalability when the channel is saturated. Moreover, to design dynamic beaconing approaches for road-safety messages, beacon rate control is the main beaconing category [10–14] in VANETs. All these application-layer approaches rely on a scalable and flexible MAC protocol to support.

For the aforementioned considerations, we propose *SS-MAC*, a novel time *S*lot-*S*haring *MAC* for efficient and scalable road-safety messages broadcasting. In *SS-MAC*, broadcast requirements of road-safety applications are periodic with different rates. It is profitable to make multiple vehicles alternately broadcast on the same time slot via inerratic coordination. To capture the periodic characteristics of road-safety applications over time slots, we first introduce a *circular recording queue* to online perceive time slots occupying status. The circular recording queue records the time slot status (occupied or vacant) during the latest K successive frames, and a suitable K recording queue can help perceive the seasonal occupied behaviors on each time slot. Based on the information, we design a distributed time slot sharing *(DTSS)* approach, to decide whether the time slot can support the sharing for a certain periodical broadcasting requirement, and how to share a time slot in an efficient way. Specifically, we present the precondition of a time slot sharing among vehicles. To satisfy the precondition, we advocate normalizing vehicle cycles for consolidated sharing. We then define the *feasibility parameter*

---

[1]For reliable transmission without collisions, vehicles in two-hop communication ranges should not use the same time slot.

[2]The priority is judged by the broadcast rate requirement in this chapter.

and *sharing potential parameter* as elaborated in Sect. 4.3 to motivate DTSS design with *perfect sharing* property. After that, we design a random index first fit *(RIFF)* algorithm based on the heuristic packing method, to conduct online vehicle-slot matching with maximizing the resource utilization of the network. We prove the efficacy of DTSS algorithm theoretically and evaluate the efficiency of RIFF algorithm by using Matlab simulations. In addition, with considering various driving scenarios and resource conditions, we conduct extensive implementation simulations to demonstrate the efficiency of *SS-MAC* in terms of *delay ratios* of the overall network and individual vehicle. The main contributions are threefold.

- We design a novel time slot-sharing MAC, referred to as *SS-MAC*, for road-safety messages broadcasting in VANETs. As *SS-MAC* supports diverse broadcasting rates, vehicles can access the medium in accordance with the underlying road-safety demands. Hence, *SS-MAC* can be scalable in terms of channel resource management, which can ably adapt to all sorts of driving scenarios.
- To achieve the common agreement of the time slot sharing, we devise a distributed time slot sharing approach, called DTSS, to regulate the time slot sharing process among vehicles. It can maximize the sharing potential of each specific time slot.
- We develop the RIFF algorithm to assist the vehicle in selecting a suitable time slot for sharing. It can not only satisfy the broadcasting requirement of vehicles but also optimize the resource utilization of the network.

The remainder of this chapter is organized as follows. The system model and preliminaries about TDMA-based MAC are introduced in Sect. 4.2. We elaborate on the design of *SS-MAC* in Sect. 4.3. We evaluate the efficiency of the RIFF algorithm and conduct extensive implementation simulations for performance evaluation of *SS-MAC* in Sect. 4.4. Section 4.5 gives a brief summary.

4.2 System Model and Preliminaries About TDMA-Based MAC

In this section, we first simply introduce the system model, and then present the preliminaries about TDMA-based MAC.

4.2.1 System Model

The system model is identical with that in *MoMAC*. As shown in Fig. 4.1, nodes communicate with each other through wireless communication. For wireless communications, the DSRC module is adopted by each node for information exchanges [15–18]. We focus on the MAC design running on CCH for road-safety applications.

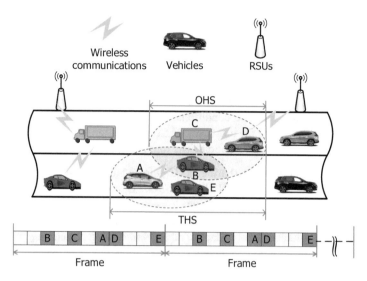

Fig. 4.1 Illustration of the system model

The medium resource of the channel is set to be a slotted/framed structure. Specifically, time is partitioned into frames, each containing N number of time slots, and each time slot has an equal-length duration for data transmission. Before the data transmission, a node has to apply for a vacant time slot. As shown in Fig. 4.1, the elements in light color denote vacant time slots, while the elements in dark color denote occupied time slots. In addition, the channel is thought to be symmetric, i.e., if the node x is in the communication range of the node y iff y is in the communication range of x. Each radio in the network has the identical communication capability with a communication range R.

For vehicles,[3] each is equipped with one DSRC radio and a GPS receiver. The GPS receiver not only provides the location information but also conducts the synchronization of time slot index among vehicles. For road safety, each vehicle is required to broadcast beacon periodically with a different rate, which is determined by the underlying road-safety application type, and all the road-safety applications have a low tolerance for delay. According to the vehicle safety communications report of the U.S. Department of Transportation, the beaconing frequency of road-safety applications ranges from 1 to 10 Hz [9]. Each vehicle is identified by a MAC address as well as a randomly generated short identifier (ID).

---

[3]As RSUs can access the channel via the same MAC protocol like vehicles, for convenience, we call the node or vehicle on behalf of the set of RSU and vehicle in this chapter.

4.2.2   Preliminaries About TDMA-Based MAC

Before transmission, each node has to acquire a unique time slot, and once a time slot is assigned, the node can use it in all subsequent frames unless a time slot collision is detected. As shown in Fig. 4.1, all the neighboring vehicles in the communication range of the vehicle A constitute its OHS. In addition, the THS of the vehicle A refers to all the vehicles that can reach it in two hops at most. Before introducing our design, in this subsection, we present some preliminaries about TDMA-based MAC,[4] which will also be adopted in *SS-MAC*.

Broadcasting Additional Frame Information   To access time slots in a distributed way, additional information exchange among neighbors is in need. In each beacon, in addition to application data, the vehicle (say vehicle i) should also broadcast the information of $I(j)$ and $T(j)$ for $\forall j \in N_{cch}(i)$, where $I(j)$ and $T(j)$ is the vehicle ID and the time slot index acquired by the vehicle j, and $N_{cch}(i)$ is the OHS of the vehicle i (including i itself). By this way, each vehicle can perceive time slot acquisitions within its THS (i.e., interference range).

Accessing Time Slots   To access a time slot, a vehicle (say vehicle k) first has to listen to the channel for N successive time slots, in order to obtain $T(j)$ for $\forall j \in N_{cch}^2(k)$, where $N_{cch}^2(k)$ is the THS of the vehicle k. After that, the vehicle can randomly choose a time slot from the free time slot set, i.e., $\mathcal{N} - T(j)$ for $\forall j \in N_{cch}^2(k)$, where $\mathcal{N}$ is the entire time slot set. Once the vehicle successfully acquires a time slot, it can use the same time slot in all subsequent frames unless a time slot collision is detected.

Detecting Time Slot Collisions   With the network topology varying, multiple vehicles may access the same time slot, leading to massive transmission collisions. To detect a time slot collision, at the end of every frame, each vehicle has to check the frame information received during previous N time slots. Specifically, for a vehicle i, if all beacons received from j for $\forall j \in N_{cch}(i)$, indicating that $i \in N_{cch}(j)$, it means no concurrent transmissions happen during the time slot $T(i)$; otherwise, the vehicle i may collide with other vehicles in its THS. Once a collision is detected by a vehicle, it has to release its original time slot and try to acquire a new time slot.

---

[4]For details, please refer to the previous chapter.

4.3 SS-MAC Design

4.3.1 Design Overview

Under a distributed and highly dynamic vehicular network, a fine-grained negotiation is in need to achieve a scalable slot-sharing TDMA-based MAC for diverse
broadcasting rates. In the following subsections, we first introduce a circular recording queue to perceive time slots occupying status. Then, we design a distributed time
slot sharing approach, called DTSS, to decide whether the time slot can support the
sharing for a specific periodical broadcasting demand, and how to share the time slot
efficiently. After that, based on the heuristic packing method, we devise an algorithm
named RIFF to conduct online vehicle-slot matching with maximizing the resource
utilization of the network.

4.3.2 Perceiving Time Slots Occupying Status

Prior to sharing a time slot, vehicles need to perceive the time slot using status.
As road-safety applications periodically broadcast beacons with different rates, one
possible solution is that each vehicle adopts a *circular recording queue* to record the
most recent using status (one means *occupied* while zero means *free*) of every slot
in each frame. Specifically, for each time slot i, $i \in [1, N]$, the vehicle initialises
a circular recording queue, denoted as $Q_i = [q_{K-1}, q_{K-2}, \ldots, q_0]$. For q_j ($j \in
[0, K-1]$), it is the $(j+1)$th element in the queue counted from right to left which
means the using status of slot i in the previous jth frame, and q_0 means the using
status in the current frame. For each vehicle, at the end of each frame, based on the
time slots using status $U(x)$ of the current frame, the status of Q_i can be updated.
Particularly, if the slot i is used by its THS neighbors in $N(x)$, elements in Q_i move
one step towards left direction (i.e., the value q_j is replaced by q_{j-1}), and the q_0 is
set to the value one; otherwise, elements in Q_i move one step towards left direction
and the q_0 is set to the value zero. With Q_i recording the most recent K frame time
slots occupying information, the sharing status of every time slot can be perceived
online, which can be the coordination guideline for vehicles to share the time slot.

To design an appropriate queue size K, the following tradeoff should be
considered. On the one hand, if the queue size is too small, the K frame time
cannot cover a complete broadcasting cycle of low-rate road-safety applications,
which means that the circular recording queue cannot provide needed coordination
information. On the other hand, if K is too large, the past K frames can last
relatively long, during which the network topology and the time slot sharing status
may change due to the high dynamics of VANETs. As a result, the outdated records
may incur inaccurate decision of time slot sharing.

4.3.3 Distributed Time Slot Sharing Approach

In this subsection, we first show the precondition of a time slot sharing among vehicles. To satisfy the precondition, we then advocate normalizing broadcast cycles for consolidated sharing. We finally elaborate the algorithm design of DTSS, in which the feasibility parameter and sharing potential parameter are defined to maximize the sharing potential of a specific time slot. Under this design, we theoretically prove that DTSS can work with the perfect sharing property.

Precondition of Time Slot Sharing As broadcast requirements for road-safety applications are periodic with different cycles, it is profitable to make multiple vehicles alternately broadcast on the same time slot, in order to achieve efficient utilization of channel resources. For convenience, we name the road-safety applications as *t-cycle* applications in this chapter, when the road-safety application has a broadcast requirement of every t frame (each frame usually lasting 100 ms).

Lemma 4.1 *For t_1-cycle and t_2-cycle applications, $1 < t_1 \leq t_2 \leq T$,[5] T is the maximum cycle value, if $t_2 = n * t_1$, $\forall n = 1, 2, 3, \ldots$, then these two applications can share a same time slot.*

Proof As shown in Fig. 4.2, the t_1-cycle application broadcasts once at the time slot $i, i \in [1, N]$, every t_1 frames. The deep color arrow is an integer pointer p_1 pointing to the frame that the t_1-cycle application will broadcast on the time slot i in this frame, $p_1 \in [0, t_1 - 1]$. For the t_2-cycle application, it also broadcasts on the time slot i, and the light color arrow is an integer pointer p_2, pointing to the frame that the t_2-cycle application will use the time slot i in this frame, for $p_2 \in [0, t_1 - 1]$ and $p_2 \neq p_1$ at the initial stage.[6] As $t_2 = n * t_1$, after $n * t_1$ frames, the t_2-cycle application will broadcast a new message at the time slot i, and the value of p_1 and

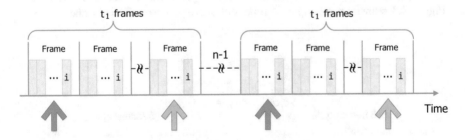

Fig. 4.2 Precondition of time slot sharing for two periodical road-safety applications

---

[5]As 1-cycle applications need to broadcast every frame, they cannot share a time slot with others.
[6]When two periodical broadcast applications share a time slot, they should use the time slot at different frames; otherwise collisions will happen.

p_2 will not change. Without collisions under this periodic activities, then the lemma is proved.   □

However, if $t_2 = n * t_1 + k, k \in [1, t_1 - 1]$, then p_2 could be $(p_2 + k * m)\%t_1$, where % is the modulus operator. As a result, p_1 has the possibility to collide with p_2, depending on the values of t_1, t_2, p_1 and p_2. We do not consider this sharing situations due to the lack of regularity.

Normalizing Cycles for Consolidated Sharing Lemma 4.1 shows a precondition of slot sharing, which also constrains the sharing conditions and brings complexity for coordination, especially in a distributed system. Particularly, for periodical applications with various cycles, few of them can have multiple relationships in terms of cycles. Moreover, for vehicles, it is hard to detect all broadcast cycles under a wide range. One reasonable solution is to normalize the cycle of applications to a close value, which not only satisfies the time requirement of road-safety applications, but also is convenient for sharing the time slot with others. To do it, we define a list of normalizing targets, denoted as $[n_1, n_2, \ldots, n_Z]$, $1 < n_1 < n_2 < \ldots < n_Z \leq T$, where n_z for $z \in [1, Z]$ is the zth target in the list, and $n_{z+1} = X * n_z$ (X is an integer and $X > 1$).

Rule 1 For a t-cycle road-safety application, $\forall t \in [1, T]$, the cycle of the application is normalized as follows

$$t = \begin{cases} 1 & t < n_1; \\ n_z & n_z \leq t < n_{z+1}; \\ n_Z & n_Z \leq t \leq T. \end{cases} \tag{4.1}$$

After normalization, periodic applications can share the time slot with each other once the time slot has enough capacity. To collect the complete coordination information, the size K of circular recording queue can be set to the value n_Z. Figure 4.3 shows an example of time slot sharing, where $Z = 3$ and $n_1 = 2$,

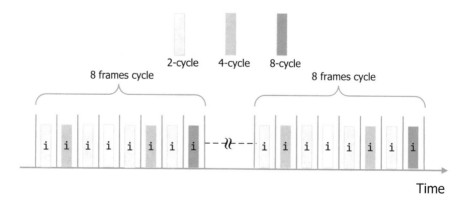

Fig. 4.3 An example of time slot sharing, where $Z = 3$ and $n_1 = 2, n_2 = 4, n_3 = 8$

$n_2 = 4$, $n_3 = 8$. In the figure, a 2-cycle application, a 4-cycle application, and a 8-cycle application are sharing a time slot i, $i \in [1, N]$. By checking the circular recording queue of the time slot i, vehicles can perceive the sharing status of the time slot, and comprehend that it can still support the sharing of another 8-cycle application.

DTSS Algorithm Design After normalizing the cycle of applications, another issue is how to conduct the coordination among vehicles to share a time slot, which should not only avoid collisions, but also can fully utilize the time slot. As the size K of circular recording queue is set to be the biggest cycle n_Z, the occupying status of K frames for a time slot can repeat in the following subsequent K frames, and vehicles can coordinate to share a time slot solely based on the recording queue information. To regulate the usage of a time slot, we design the DTSS algorithm.

Definition 4.1 (Item Size) For a t-cycle road-safety application, the *item size* of the application, is denoted by α with $\alpha = \frac{1}{t}$, $\alpha \in (0, 1]$.

Definition 4.2 (Slot Capacity) For a specific time slot, the *capacity* C of the time slot is calculated by the number of idle elements in the circular recording queue to the total size K, and the capacity of a free time slot is $C = 1$.

Remark 4.1 For a road-safety application with size α, if the application has a chance to share the time slot, iff the remain capacity of the time slot is equal or larger than the size α, i.e.,

$$C \geq \alpha. \tag{4.2}$$

For a t-cycle application, after normalizing the cycle to a value n_z, we define a list of serial numbers for the recording queue under this cycle by modulus operator. Specifically, for an element q_j in the queue, $j \in [0, K - 1]$, the according serial number $s_j = j \% n_z$, $s_j \in [0, n_z - 1]$, is defined.

Definition 4.3 (Feasibility Parameter) We define the *feasibility parameter* f_j^z for each element in the circular recording queue under different cycle values, $j \in [0, K - 1]$ and $z \in [1, Z]$. Given z and j, its serial number satisfies $s_j = j \% n_z$. For $\forall \{x | x \% n_z = s_j, x \in [0, K - 1]\}$, if all $q_x = 0$, then f_j^z is set to the value one, otherwise it is set to the value zero, i.e.,

$$f_j^z = \begin{cases} 1, \forall \{x | x \% n_z = s_j, x \in [0, K - 1]\}, q_x = 0; \\ 0, elsewhere. \end{cases} \tag{4.3}$$

Remark 4.2 If a n_z-cycle application can share a time slot, iff there is a $j \in [0, n_z - 1]$ and $f_j^z = 1$, i.e.,

$$\exists j \in [0, n_z - 1], s.t., f_j^z = 1. \tag{4.4}$$

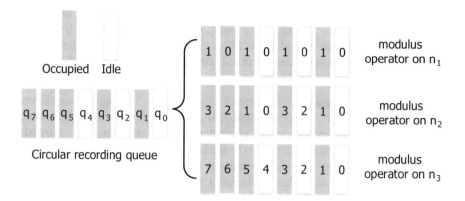

Fig. 4.4 An example of a circular recording queue and the corresponding modulus operators on different cycle values, where $K = 8$, and $n_1 = 2, n_2 = 4, n_3 = 8$

If the application chooses the jth element in the recording queue to share the time slot, then all the elements $\forall \{x | x \% n_z = s_j, x \in [0, K - 1]\}$ will be occupied by the application to meet its requirement. Following Remarks 4.1 and 4.2, vehicles can share a time slot without collisions. However, choosing an appropriate value j in Remark 4.2 determines the sharing efficiency of the time slot. For instance, Fig. 4.4 shows an example of the circular recording queue of a specific time slot and the corresponding modulus operators on different cycle values, where the size of recording queue K is set to be 8 and $n_1 = 2, n_2 = 4, n_3 = 8$ respectively. In addition, the dark block denotes the occupied status while the white block denotes the idle status and the remain capacity of the time slot is calculated by the number of idle blocks to the total size K, i.e, $\frac{3}{8}$. According to the capacity restriction of Remark 4.1, only the 4-cycle or 8-cycle applications can be supported by this time slot. For an 8-cycle application, the element in the recording queue with a serial number 0, 2, 4 can satisfy its demand. If the application chooses to occupy the serial number 2, then the remaining capacity of the time slot can still support a 4-cycle application, otherwise the time slot can only support 8-cycle applications if the number 0 or 4 is chosen.

Remark 4.3 For the jth element in the recording queue, if $f_j^z = 0$, then $f_j^{z-1} = 0$, $z \in [2, Z]$.

Remark 4.3 indicates that, if the jth element cannot support sharing for the n_z-cycle applications, then it has no potential for supporting applications with a higher frequency. Considering this, we define the *sharing potential parameter* p_j for each element in the recording queue as follows.

Definition 4.4 (Sharing Potential Parameter) For the jth element in the recording queue, the sharing potential parameter p_j is the maximum item size of the n_z-cycle applications that the corresponding value of feasibility parameter satisfies

$f_j^z = 1$, i.e.,

$$p_j = \frac{1}{n_z}, z = \min\{\forall z \in [1, Z], s.t., f_j^z = 1\}. \tag{4.5}$$

In addition, the sharing potential of the time slot p is the maximum p_j, for $j \in [0, K - 1]$, i.e.,

$$p = \max\{p_j, j \in [0, K - 1]\}. \tag{4.6}$$

Remark 4.4 If an n_z-cycle application can share a time slot, iff the sharing potential of the time slot p satisfies $p \geq \frac{1}{n_z}$.

Rule 2 For an n_z-cycle application, if a list J of elements in the recording queue($j \in [0, n_z - 1]$) can satisfy its sharing demand according to the limit of Remarks 4.1 and 4.2, DTSS algorithm chooses the element with the minimum value of sharing potential parameter from the list J.

As shown in Fig. 4.4, for an 8-cycle application, the 0th, 2th and 4th element can satisfy its requirement with the sharing potential parameter $p_0 = \frac{1}{4}$, $p_2 = \frac{1}{8}$, $p_4 = \frac{1}{4}$ respectively. According to Rule 2, the 2th element will be chosen and the application can access the channel at this time slot after waiting for $(n_z - 1 - j)$ frames with a repeating cycle of 8 frames. DTSS algorithm for distributed and efficient time slot sharing is shown in Algorithm 1.

Algorithm 1 DTSS algorithm for a time slot sharing

Input: Q_i and cycle t
Output: waiting frames w
1: Initialize: $w = -1$, $p_{min} = 1$
2: Normalize: $t, t \leftarrow n_z$
3: **if** $C_i \geq \alpha_z$ **then**
4:     **for** $j \in [0, n_z)$ **do**
5:         **if** $f_j^z == 1$ **then**
6:             **if** $p_j \leq p_{min}$ **then**
7:                 $p_{min} = p_j$
8:                 $w = n_z - 1 - j$
9: **return** w

Definition 4.5 (Perfect Sharing) We define the *perfect sharing* property for a time slot, if the time slot has the following feature. For a time slot with the capacity C and a application with the normalized cycle $n_z (z \in [1, Z])$, if $C \geq \frac{1}{n_z}$, then the time slot can support sharing for the application.

Lemma 4.2 *DTSS algorithm can guarantee each time slot with the perfect sharing property.*[7]

Proof For a time slot, it can support sharing for n_1 numbers of n_1-cycle applications, and an n_z-cycle application can be divided into X groups of elements for X n_{z+1}-cycle applications sharing, i.e., $n_{z+1} = X * n_z$. Considering a time slot is being shared by the n_z-cycle application with the number of x_z, $x_z \geq 0$, $z \in [1, Z]$ and

$$\sum_{z=1}^{Z} x_z \frac{1}{n_z} \leq 1. \tag{4.7}$$

According to Rule 2 of choosing the minimum sharing potential element, it can guarantee that if the group elements for a n_z-cycle application still have the space for an n_{z+1} application, the n_{z+1} application will choose elements from the remaining and will not occupy elements which have potentials for other n_z-cycle applications. Based on this rule, the elements occupied by X numbers of n_{z+1}-cycle applications finally can be combined to a group of elements for an n_z-cycle application. Then the occupying status of a time slot can be represented by the n_i-cycle application with the number of y_i, $0 \leq y_i < X(y_1 < n_1)$, $i \in [1, Z]$. Under this considering, the capacity of the time slot $C = \sum_{i=1}^{Z} l_i \frac{1}{n_i}$ and the l_i satisfies

$$l_i = \begin{cases} n_1 - 1 - y_1 & i = 1; \\ X - 1 - y_i & 1 < i < Z; \\ X - y_i & i = Z. \end{cases} \tag{4.8}$$

To support the requirement of an n_j-cycle application, if $\exists i \in [1, j]$, $s.t.$, $l_i > 0$, apparently the time slot can support sharing for the application; if $\forall i \in [1, j]$, $l_i = 0$, according to the expression of the capacity C and the limit of Eq. (4.8), thus the $C < \frac{1}{n_j}$ which violates the condition in Definition 4.5. Then the lemma is proved.

□

4.3.4 Online Vehicle-Slot Matching Approach

Given the recording queue information of a time slot, DTSS algorithm can determine whether the time slot can satisfy the broadcast requirement of applications and how to share a time slot greedily to maximize future sharing potential. However, in practice, the medium is set with various time slots, and how to select a satisfying time slot for nodes can significantly affect the network resource utilization. According to Remark 4.4, each time slot has a sharing potential value,

---

[7]Lemma 4.2 demonstrates the efficacy of the DTSS algorithm design in theory.

and only when the item size of the application is smaller than the potential value, then the application can adopt the time slot. Inspired by this, the online vehicle-slot matching problem can be modeled as an online bin packing problem, where each application has an item size and each free time slot has a full potential 1. The online bin packing problem is a well-known NP-hard problem, and some classical online algorithms are proposed, such as WF (Worst-Fit), BF (Best-Fit), FF (First-Fit) and so on. These classical algorithms can be generalized to the *Any-Fit* and *Almost Any-Fit* classes [19].

Any-Fit Constraint If $B_1, B_2, \ldots\ldots B_j$ are the current nonempty bins, the current item will be packed into B_{j+1} iff it does not fit in any of the bins $B_1, B_2, \ldots\ldots B_j$.

Almost Any-Fit Constraint If $B_1, B_2, \ldots\ldots B_j$ are the current nonempty bins and the $B_k(k \leq j)$ is the unique bin with the smallest content, the current item will be packed into B_k iff it does not fit in any of the bins to the left of B_k.

The class of online heuristics that satisfies the Any-Fit constraint will be denoted by $\mathcal{AF}$, and the class of online algorithms satisfying both constraints above will be denoted by $\mathcal{AAF}$.

Theorem 4.1 (Johnson [20]) *For every algorithm $A \in \mathcal{AF}$,*

$$R_{FF}^{\infty} = R_{BF}^{\infty} \leq R_A^{\infty} \leq R_{WF}^{\infty}, \tag{4.9}$$

where R_A^{∞} is the asymptotic worst-case ratio (or asymptotic performance ratio), and the number is the value of packings produced by algorithm A compared to the optimal packings in the worst case.

Theorem 4.2 (Johnson [20]) *For every algorithm $A \in \mathcal{AAF}$,*

$$R_A^{\infty} = R_{FF}^{\infty}. \tag{4.10}$$

Theorems 4.1 and 4.2 demonstrate that the FF and BF algorithms can achieve the lowest worst-case ratio compared with algorithms in $\mathcal{AF}$ and $\mathcal{AAF}$ class. In FF algorithm, the current item will be packed into the first nonempty bin which it fits, and if no such nonempty exists, the algorithm will open a new bin to pack the item. Differently, the BF algorithm packs the current item into an open bin with largest content which it fits, and if no such nonempty exists, the algorithm will open a new bin. However, during time slot acquisitions, there are many nodes may acquire time slots simultaneously. The fitting results for different nodes in FF or BF, are very likely to be the same value, which can result in collisions when more than two nodes are assigned to the same time slot. To cope with it, based on FF, we design a new heuristic algorithm named *RIFF*. If $\mathcal{U} = \{U_1, U_2, \ldots\ldots, U_j\}$ is the current used time slot list and $U_i(i \in [1, j])$ is the ith element in the list $\mathcal{U}$, RIFF algorithm firstly generates a random variable $k(k \in [1, j])$ to index the elements in the $\mathcal{U}$ as follows

$$I = \begin{cases} i + j - k + 1 & i < k; \\ i - k + 1 & i \geq k, \end{cases} \qquad (4.11)$$

where the value I is the index of the element. RIFF algorithm then allocates the lowest indexed time slot which can satisfy the broadcast requirement of the application to the current node; if no such used time slot exists, RIFF will randomly choose a free time slot for the node. Apparently, RIFF algorithm satisfies the Any-Fit constraint and meets the demand of the Almost Any-Fit constraint at the most of time. Algorithm 2 shows the pseudocode for the RIFF algorithm.

Algorithm 2 RIFF algorithm for online vehicle-slot matching

Input: $\mathcal{U}$, S and cycle t
Output: time slot s and waiting frames w
1: Initialize: $s = 0, w = -1$
2: **if** $t == 1$ **then**
3:      $s \leftarrow random(S - \mathcal{U})$
4:      $w = 0$
5: **else**
6:      $k = random(1, len(\mathcal{U}))$
7:      **for** $i \in (1, len(\mathcal{U}))$ **do**
8:          $index = i$
9:          **if** $index < k$ **then**
10:             $index = index + len(\mathcal{U}) - k + 1$
11:         **else**
12:             $index = index - k + 1$
13:         $w = DTSS(Q_{\mathcal{U}_{index}}, t)$
14:         **if** $w \neq -1$ **then**
15:             $s = \mathcal{U}_{index}$
16:             **break**
17:     **if** $s == 0$ **then**
18:         $s \leftarrow random(S - \mathcal{U})$
19:         $w = 0$
20: **return** s, w

4.4   Performance Evaluation

In this section, we perform the performance evaluation, where the performance of RIFF algorithm is evaluated based on Matlab simulations, and implementation simulations are carried out to evaluate the efficiency of *SS-MAC*.

4.4.1 Evaluation of RIFF Algorithm

Since we have theoretically proved the efficiency of DTSS algorithm in the Sect. 4.3.3, in this subsection, only the proposed online vehicle-slot matching algorithm will be evaluated. The following two benchmarks are designed to compare the performance with the RIFF algorithm.

- **Random Fit Approach**. The Random Fit approach allocates the time slot randomly once the time slot can satisfy the demand of the vehicles.
- **FF Approach**. In this approach, the current vehicle will be associated with the lowest nonempty indexed time slot, which can satisfy its sharing demand. If there is no such a nonempty time slot, the current vehicle will be associated with a new free time slot.

The two benchmarks are reasonable approaches that would be widely employed in practice, since it is easy to implement the Random Fit approach, and it has been proved that among all online matching heuristics, the FF approach can achieve the best performance in terms of worst-case properties. To evaluate the performance, for all candidate matching approaches, vehicles will adopt the DTSS algorithm to share a specific time slot after the matching process. For parameter setup, we configure each frame with 100 time slots, and set the application cycle t of each vehicle randomly within the range between 1 and 10, where the unit is the time duration equaling one frame. In each round simulation, the matching algorithms will assign the time slot for a new vehicle. Table 4.1 shows the detailed simulation parameters. The following two metrics are designed to evaluate the performance of candidate matching approaches.

- **Number of occupied time slots**: refers to the required time slots to satisfy the broadcasting demands of matched vehicles.
- **Sharing efficiency**: refers to the ratio of the sharing capacity to the total capacity of a time slot.

We plot the number of occupied time slots during the matching process with different matching algorithms in Fig. 4.5, where we include the theoretical result[8]

Table 4.1 Evaluation parameters for RIFF algorithm

Parameters	Value
Number of time slots	[1, 100]
Cycles of applications (in *frames*)	[1, 10]
Number of vehicles	[1, 1000]
Matching rounds	[1, 500]
Simulation times	50

---

[8]The theoretical result is obtained by calculating the sum sizes of all matched vehicles, which cannot be achieved in practise.

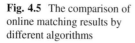

Fig. 4.5 The comparison of online matching results by different algorithms

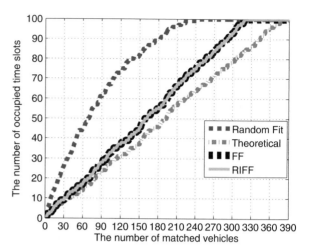

to show the performance upper bound. With the results, the following three major statements can be made. First, the algorithms of RIFF and FF can achieve the similar performance with the two curves intertwining closely. Second, the algorithm of RIFF can outperform the Random Fit algorithm significantly when medium resources are sufficient. For example, when assigning time slots for 100 vehicles, only 30 time slots are required to support vehicles with the RIFF algorithm, but with the Random Fit algorithm, 66 time slots are required for data transmission. It means that more than half of resources are saved, which can significantly enhance the resource utilization. Besides, our proposed RIFF algorithm can achieve the near-optimal performance compared with the theoretical result, which requires 26 time slots, being close to the value of 30. Third, compared with the Random Fit algorithm, our proposed RIFF algorithm can serve much more vehicles when meeting the resource shortage condition. For instance, when with 100 time slots, the Random Fit algorithm can support 240 vehicles, but the RIFF algorithm can serve 320 vehicles under the same resource condition.

In Fig. 4.6, we plot the average sharing efficiency of time slots with different algorithms during the matching process. Note that, the curve of FF algorithm is omitted since its performance is tightly close to that of the RIFF algorithm, which may blur the figure and confuse the presentation. The following two major observations can be obtained. First, the RIFF algorithm can always achieve a better performance with higher average sharing efficiencies when compared with the Random Fit algorithm. Second, when adopting the RIFF algorithm, most sharing efficiencies can reach 98% during the matching process, which can drop down to 50% when adopting the algorithm of Random Fit, which is a significant performance gap. This can well explain why our proposed RIFF algorithm can enhance the resource utilization significantly when compared with the Random Fit algorithm.

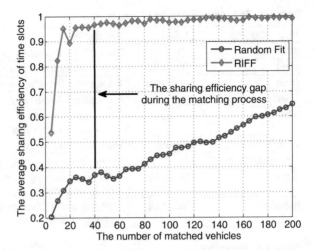

Fig. 4.6 The sharing efficiency comparison between the RIFF and the Random Fit algorithm

4.4.2   Evaluation of SS-MAC

In this subsection, to demonstrate the efficiency of *SS-MAC*, we conduct extensive implementation simulations under various road topologies and resource conditions.

Simulation Setup To emulate realistic driving scenarios, we adopt the Simulation of Urban Mobility (SUMO) to build the transportation system [21], where two typical VANET scenarios are created, i.e., highways and urban roads. Specifically, for the highway scenario, we build a bidirectional 8-lane road segment with a length of 10 km, where each direction has four lanes and we set the speed limit to be 60 km/h, 80 km/h, 100 km/h, and 120 km/h, respectively. For the urban scenario, we build four bidirectional 6-lane road segments, each lasting 4 km long, and they converge at an intersection. For each three lanes in one direction, we set the speed limit to be 50 km/h, 60 km/h, and 70 km/h, respectively. In addition, we configure traffic lights at each inbound road segment of the intersection, where the green light duration is set to be 20 s.

Under both driving environments, we configure vehicles with different running parameters in terms of acceleration ability, deceleration ability, and maximum velocity. We set ten settings of running parameters per the main types of vehicles on the market. In addition, vehicles are driven under the LC2013 lane-changing and Krauss car-following model, to conduct necessary lane-change and acceleration/deceleration activities. Besides, the system also includes the driver imperfection and impatience parameters to mimic the human factors. We present the detailed parameters of simulated vehicles in Table 4.2.

We generate vehicles at the entrance of each road, where the urban environment has 3 × 4 entrance lanes and the highway environment has 4 × 2 entrance lanes. To mimic the normal traffic conditions, vehicles are generated with different rates, i.e., urban: 6 vehicles/lane/min, and highway: 10 vehicles/lane/min. When vehicles are generated, they will randomly choose a running parameter setting, as well

Table 4.2 Parameters of vehicles in simulations

Parameters	Values	Description
maxSpeed	[80, 240]	The vehicle's maximum velocity (in km/h).
accel	[1.0, 5.0]	The acceleration ability of vehicles (in m/s^2).
decel	[3.0, 10.0]	The deceleration ability of vehicles (in m/s^2).
length	[4.0, 7.0]	The vehicle's length (in m).
minGap	[3.0, 10.0]	The minimum offset to the leading vehicle when standing in a jam (in m).
car-following model	$Krauss$	The model used for car following.
lane-changing model	$LC2013$	The model used for changing lanes.
sigma	[0.5, 1.0]	The car-following model parameter defining the driver imperfection (between 0 and 1).
impatience	[0.5, 1.0]	Willingness of drivers to impede vehicles with higher priority (between 0 and 1).

Table 4.3 Simulation parameters

Parameters	Highway	Urban
Road length	10 km	4 km
Number of road segments	1	4
Number of intersections	0	1
Number of lanes on each road	8	6
Speed limit in lanes (in km/h)	[60, 120]	[50, 70]
Transmission range	300 m	300 m
Frame duration	100 ms	100 ms
Cycles (in frames)	[1, 10]	[1, 10]
Slot duration	1 ms	1 ms
Number of slots	(40, 50, 60)	(80, 90, 100)
Loaded vehicles	1630	1360
Running vehicles	400–500	500–600
Simulation time	1000 s	1000 s

as a destination (i.e., the end of road segment). After the vehicles reaching the destination, they will disappear from the system. For communication parameters, we set the transmission range R to be 300 m per the observation that 802.11p-compatible onboard units can support reliable data transmission within 300 m [18]. We set the frame duration to be 100 ms, in order to satisfy the highest frequency requirement of road-safety applications [9]. In addition, we set the application cycle t (in frames) within the range between 1 and 10. The detailed simulation parameters[9] are shown in Table 4.3.

Methodology The following two candidate MAC protocols are designed to compare the performance with *SS-MAC*.

---

[9]Note that, the numbers of running vehicles are recorded at the saturated traffic condition, and the number can be larger in the urban environment due to the traffic lights.

- **Aggressive MAC protocol**. In this protocol, once a vehicle has successfully acquired a time slot, no matter what the application cycle is, it will broadcast at the time slot every frame. To this end, vehicles within the same THS would perceive the time slot to be busy, and will not try to apply for it.
- **Conservative MAC protocol**. In this protocol, on the contrary, each vehicle just broadcasts when in need in accordance with the application cycle, and when the time slot is not used, other vehicles would contend for the time slot. The advantage of this protocol is that only necessary resources are used.

For performance evaluation, the metric of **delay ratio** is considered, which refers to the ratio of number of unsuccessful transmissions to the total number of needed transmissions per the cycle of applications. There are two events that could result in an unsuccessful transmission, one is the medium access collision during the transmission, and the other is the medium access failure caused by the unfairness of resource allocation. With adopting the delay ratio, how efficient the road-safety beacons are successfully delivered can be evaluated.

Performance Comparison In this subsection, performance with different MAC protocols are compared under urban and highway environments. To evaluate the performance of *SS-MAC*, we consider the delay ratio of both the individual vehicle and overall network.

We plot the cumulative distribution functions (CDFs) of delay ratio of the overall network during each second in Fig. 4.7, where different resource conditions are simulated under the highway environment. The following three major observations can be obtained with the results. First, under all resource conditions, *SS-MAC* can outperform the candidate MAC protocols significantly. For instance, when adopting *SS-MAC*, more than 80% delay ratios are smaller than 0.2 when with $N = 40$, which however can reach over 0.43 and 0.7 in respective Aggressive MAC and Conservative MAC protocol. Second, when with more resources, even though all MAC protocols can enhance the performance to achieve lower delay ratios, there is no change to the bending feature of their performance curves. Particularly, the curves of two candidate MAC protocols are *concave* with the fast growth rate of y locating at the larger x parts. In contrast, the curve of *SS-MAC* is *convex* with the fast growth rate of y locating at the smaller x parts. It demonstrates that there is always a significant performance gap between *SS-MAC* and the other two candidate MAC protocols. For example, for the delay ratio being smaller than 0.1, when adopting our proposed *SS-MAC*, the proportion can be 40%, 80%, and 90% when with $N = 40$, $N = 50$, and $N = 60$, respectively. However, the proportion is just about 2%, 15%, and 40%, respectively, when adopting the Aggressive MAC protocol, and is 3%, 10%, and 30%, respectively, when adopting the Conservative MAC protocol. The *inherent* efficiency of *SS-MAC* in comparison with candidate MAC protocols can be demonstrated by the curve concavity and convexity property. Third, when with more resources, the CDF gaps between *SS-MAC* and the other two MAC protocols can decrease. It verifies that when encountering the resource shortage condition, our proposed *SS-MAC* can work more robustly.

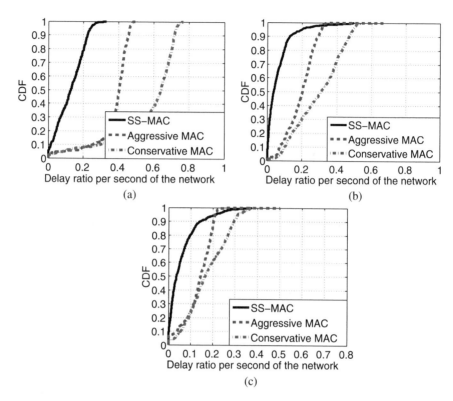

Fig. 4.7 CDFs of delay ratio of the network during each second with different resource conditions in the highway environment. (**a**) With $N = 40$. (**b**) With $N = 50$. (**c**) With $N = 60$

We plot the CDFs of delay ratio of individual vehicle in Fig. 4.8, where different resource conditions are simulated under the highway environment. Under all resource conditions, it can be easily seen that *SS-MAC* can achieve the best performance with the lowest delay ratio. For example, by adopting *SS-MAC*, the protocol can guarantee 90% vehicles with the delay ratio being smaller than 0.3 with $N = 40$. However, when adopting the Aggressive MAC and Conservative MAC protocol, the value would be increased to 0.66 and 0.88, respectively. On the other hand, under all resource conditions, the curves of *SS-MAC* keep steep, which however become flat with other two candidate MAC protocols. For instance, when adopting *SS-MAC*, to reach the proportion of 100%, the delay ratios can range from 0 to about 0.68, 0.5, and 0.41 when under respective resource conditions. However, when adopting the Aggressive MAC protocol, the delay ratios can range from 0 to respective 1, 1, and 0.74 when under different resource conditions, and the value can range from 0 to 0.97, 0.9, and 0.72, respectively, when adopting the Conservative MAC protocol. The flat curves are indicative of unfair medium access.

Similar observations are also held in urban environments. With the space limitation, we just plot the CDFs of delay ratio of individual vehicle in Fig. 4.9.

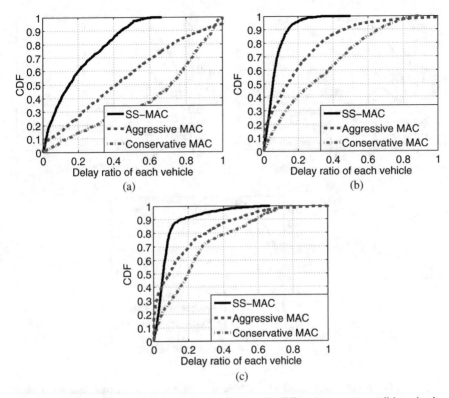

Fig. 4.8 CDFs of delay ratio of individual vehicle with different resource conditions in the highway environment. (**a**) With $N = 40$. (**b**) With $N = 50$. (**c**) With $N = 60$

With the results, it can be easily seen that under all resource conditions, *SS-MAC* can achieve the supreme performance with the lowest delay ratios.

4.5 Summary

In this chapter, we have proposed a novel time slot-sharing MAC, named *SS-MAC*, to support diverse beaconing rates for road-safety applications in VANETs. In specific, we have first introduced a circular recording queue to perceive occupancy states of time slots in real time, and then devised a distributed time slot sharing approach called DTSS to share a specific time slot efficiently. In addition, we have developed the random index first fit algorithm, named RIFF, to assist vehicles in selecting a suitable time slot for sharing with maximizing the resource utilization of the network. We have theoretically proved the efficacy of DTSS algorithm and evaluated the efficiency of RIFF algorithm by using Matlab simulations. Finally, under various driving scenarios and resource conditions, we have conducted

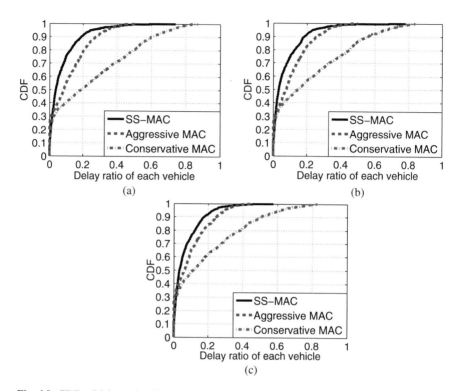

Fig. 4.9 CDFs of delay ratio of individual vehicle under different resource conditions in the urban environment. (**a**) With $N = 80$. (**b**) With $N = 90$. (**c**) With $N = 100$

extensive implementation simulations to demonstrate the efficiency of *SS-MAC*. Particularly, under both highway and urban environments, delay ratios of the overall system and individual vehicle can be significantly reduced with all sorts of resource conditions. Note that, in the previous chapter, we have proposed *MoMAC* to cope with the time slot collisions caused by vehicular mobilities. In this chapter, *SS-MAC* has been proposed to enhance the MAC scalability in supporting diverse beaconing rates with efficient resource utilization. They can work collectively to provide collision-free/reliable, scalable, and efficient medium access for moving and distributed vehicles.

References

1. M. Hadded, P. Muhlethaler, A. Laouiti, R. Zagrouba, L.A. Saidane, TDMA-based MAC protocols for vehicular ad hoc networks: a survey, qualitative analysis, and open research issues. IEEE Commun. Surv. Tutor. **17**(4), 2461–2492 (2015)
2. H.A. Omar, W. Zhuang, L. Li, VeMAC: a TDMA-based MAC protocol for reliable broadcast in VANETs. IEEE Trans. Mob. Comput. **12**(9), 1724–1736 (2013)

3. F. Borgonovo, A. Capone, M. Cesana, L. Fratta, ADHOC MAC: new MAC architecture for ad hoc networks providing efficient and reliable point-to-point and broadcast services. Wirel. Netw. **10**(4), 359–366 (2004)

4. F. Lyu, H. Zhu, H. Zhou, L. Qian, W. Xu, M. Li, X. Shen, MoMAC: mobility-aware and collision-avoidance MAC for safety applications in VANETs. IEEE Trans. Veh. Technol. **67**(11), 10590–10602 (2018)

5. X. Jiang, D.H.C. Du, PTMAC: a prediction-based TDMA MAC protocol for reducing packet collisions in VANET. IEEE Trans. Veh. Technol. **65**(11), 9209–9223 (2016)

6. W. Zhuang, Q. Ye, F. Lyu, N. Cheng, J. Ren, SDN/NFV-empowered future IoV with enhanced communication, computing, and caching. Proc. IEEE **108**(2), 274–291 (2020)

7. H. Zhou, N. Cheng, Q. Yu, X. Shen, D. Shan, F. Bai, Toward multi-radio vehicular data piping for dynamic DSRC/TVWS spectrum sharing. IEEE J. Sel. Areas Commun. **34**(10), 2575–2588 (2016)

8. N. Cheng, F. Lyu, W. Quan, C. Zhou, H. He, W. Shi, X. Shen, Space/aerial-assisted computing offloading for IoT applications: a learning-based approach. IEEE J. Sel. Areas Commun. **37**(5), 1117–1129 (2019)

9. CAMP Vehicle Safety Communications Consortium and others, Vehicle safety communications project: Task 3 final report: identify intelligent vehicle safety applications enabled by DSRC, in *National Highway Traffic Safety Administration, US Department of Transportation*, Washington, DC, March 2005

10. S.A.A. Shah, E. Ahmed, F. Xia, A. Karim, M. Shiraz, R.M. Noor, Adaptive beaconing approaches for vehicular ad hoc networks: a survey. IEEE Syst. J. **PP**(99), 1–15 (2016)

11. H. Peng, Q. Ye, X. Shen, Spectrum management for multi-access edge computing in autonomous vehicular networks. IEEE Trans. Intell. Transport. Syst. 1–12 (2019). https://doi.org/10.1109/TITS.2019.2922656

12. N. Wisitpongphan, O.K. Tonguz, J.S. Parikh, P. Mudalige, F. Bai, V. Sadekar, Broadcast storm mitigation techniques in vehicular ad hoc networks. IEEE Wirel. Commun. **14**(6), 84–94 (2007)

13. F. Lyu, N. Cheng, H. Zhu, H. Zhou, W. Xu, M. Li, X. Shen, Towards rear-end collision avoidance: adaptive beaconing for connected vehicles. IEEE Trans. Intell. Transp. Syst. 1–16. https://doi.org/10.1109/TITS.2020.2966586. Early Access, Jan. 2020

14. Y.-C. Tseng, S.-Y. Ni, Y.-S. Chen, J.-P. Sheu, The broadcast storm problem in a mobile ad hoc network. Wirel. Netw. **8**(2/3), 153–167 (2002)

15. H. Zhou, W. Xu, J. Chen, W. Wang, Evolutionary V2X technologies toward the internet of vehicles: challenges and opportunities. Proc. IEEE **108**(2), 308–323 (2020)

16. K. Abboud, H.A. Omar, W. Zhuang, Interworking of DSRC and cellular network technologies for V2X communications: a survey. IEEE Trans. Veh. Technol. **65**(12), 9457–9470 (2016)

17. H. Peng, D. Li, K. Abboud, H. Zhou, H. Zhao, W. Zhuang, X. Shen, Performance analysis of IEEE 802.11p DCF for multiplatooning communications with autonomous vehicles. IEEE Trans. Veh. Technol. **66**(3), 2485–2498 (2017)

18. F. Lyu, H. Zhu, N. Cheng, H. Zhou, W. Xu, M. Li, X. Shen, Characterizing urban vehicle-to-vehicle communications for reliable safety applications. IEEE Trans. Intell. Transp. Syst. 1–17. https://doi.org/10.1109/TITS.2019.2920813. Early Access, Jun. 2019

19. E.G. Coffman Jr, J. Csirik, G. Galambos, S. Martello, D. Vigo, Bin packing approximation algorithms: survey and classification, in *Handbook of Combinatorial Optimization* (Springer, New York, 2013), pp. 455–531

20. D.S. Johnson, Fast algorithms for bin packing. J. Comput. Syst. Sci. **8**(3), 272–314 (1974)

21. DLR Institute of Transportation Systems, Sumo: Simulation of urban mobility. http://www.dlr.de/ts/en/desktopdefault.aspx/tabid-1213/

Chapter 5
Characterizing Urban V2V Link Communications

After enhancing the communication performance at the MAC layer, in this chapter, we investigate the vehicular link layer performance, which is of paramount importance for reliable message exchanging. However, with limited literature available, there is a lack of understanding about how IEEE 802.11p based DSRC performs for V2V communications in urban environments. In this chapter, we conduct intensive data analytics on V2V communication performance, based on the field measurement data collected from off-the-shelf IEEE 802.11p-compatible onboard units (OBUs) in Shanghai city, and obtain several key insights as follows. First, among many context factors, non-line-of-sight (NLoS) link condition is the major factor degrading V2V performance. Second, both line-of-sight (LoS) and NLoS durations follow power law distributions, which implies that the probability of having long LoS/NLoS conditions can be relatively high. Third, the packet inter-reception (PIR) time distribution follows an exponential distribution in LoS conditions but a power law in NLoS conditions. In contrast, the packet inter-loss (PIL) time distribution in LoS condition follows a power law but an exponential in NLoS condition. Fourth, the overall PIR time distribution is a mix of exponential distribution and power law distribution. The presented results can provide solid ground to validate models, tune VANET simulators, and improve communication strategies.

5.1 Problem Statement

IEEE 802.11p-based DSRC [1–5] has been a standard, customized for severe-fading and highly mobile vehicular environments. Based on DSRC, V2X[1] communications become the essential component to enable cooperative road-safety applications [6–

---

[1]V2X (i.e., vehicle-to-everything) means the wireless communication between vehicle and any other entity on road, which broadly includes communication paradigms of V2V, V2I, V2P, etc.

© Springer Nature Switzerland AG 2020
F. Lyu et al., *Vehicular Networking for Road Safety*, Wireless Networks, https://doi.org/10.1007/978-3-030-51229-3_5

9]. Understanding the characteristics of 802.11p-based DSRC, especially in urban environments, is quite important for vehicular network protocols and road-safety applications.

However, to characterize the behavior of V2V[2] link performance in urban environments, is very challenging due to the following three reasons. First, as urban environments are complex and highly dynamic, too many uncontrollable factors, such as time-varying traffic conditions, various types of roads, and all different surrounding trees and buildings [10–13], can affect V2V link performance. It is hard to separate the impact of each factor on the final DSRC link performance. Second, to conduct realistic studies on urban V2V communications, experiments should involve different traffic conditions, road types, and cover a sufficiently long time, which are labor-intensive and time-consuming. The lack of real-world trace is the hurdle of achieving efficient protocol design and precise model development [14]. Third, to thoroughly capture the link variation in the moving, various metrics should be comprehensively investigated. Performance analytics with single or limited metrics not only may give one-sided communication knowledge, but also can confuse researchers and application designers without providing multi-perspective clues.

In the literature, some measurement-based DSRC studies have been carried out. Meireles et al. [15] and M. Boban et al. [16] focused on investigating the impact of obstacles between the communication link. They confirmed that line-of-sight (LoS) and non-line-of-sight (NLoS) conditions could deeply affect the DSRC performance, based on which, they then designed V2V propagation models with taking LoS and NLoS conditions into consideration. However, they conducted experiments by fixing two communicating vehicles, and did not investigate further with moving vehicles. Similarly, physical layer measurements on DSRC channels are conducted [17–20], in which the characteristics of the path loss, coherence time, Doppler spectrum, etc., were investigated. All these findings could be very different when vehicles move. In this chapter, we do not model a LoS/NLoS channel but focus on investigating their impacts on V2V communications. In addition, we pay little attention to physical layer features, as they vary dramatically with vehicle moving and are impossibly characterized in patterns. By collecting communication trace from moving vehicles, Bai et al. [21] investigated the metric of packet delivery ratio (PDR), and Martelli et al. [22] studied the metric of packet inter-reception (PIR) time, which refers to the probability of successfully receiving a packet, and the interval of time elapsed between two successfully received packets, respectively. However, in both pieces of work, very limited metrics are evaluated. Besides, they did not discriminate between different channel conditions in terms of LoS and NLoS, and drew their conclusions based on all aggregated measurements, which may bias from the ground truth. Nevertheless, there is no statistical study on the

---

[2]In this chapter, we concentrate on V2V performance since it is highly related to the road safety due to the fast speed of moving vehicles.

impact of channel conditions in terms of LoS and NLoS, and how these two conditions interact when vehicles move in urban environments.

In this chapter, we conduct an empirical study on 802.11p-based V2V communications in urban environments. As large volumes of real-world data are essential to realistic analysis, we implement a V2V communication testbed consisting of two experimental vehicles, each equipped with an 802.11p-compatible onboard unit (OBU), one GPS receiver, and two tape records. With the testbed, V2V beaconing data and the simultaneous environmental context information can be collected. We conduct an extensive data collection campaign in three typical environments in Shanghai city, i.e., urban, suburban, and highway, which lasts for more than 2 months and covers a total distance of over 1500 km. Moreover, with the whole campaign typed, we visually label out all LoS and NLoS conditions for all traces of different urban environments. With these valuable traces, we first analyze PDR across all traces and have the following two major observations. *First, "perfect zone" (i.e., the portion of PDR larger than 80%) prevails throughout a wide communication range (e.g., 300 m in our case) in urban vehicular networks*, which implies 802.11p V2V communication is particularly reliable across all urban environments. This observation is very unlike the "gray-zone phenomenon" reported in work [21]. *Second, within a rather long range (e.g., 500 m in our case), it is the NLoS conditions instead of long distances that affect PDR the most.* The probability that a pair of vehicles are blocked by obstacles, such as in-lane traffic and slopes, increases as the distance between this pair of vehicles increases.

Given the importance of NLoS conditions, we then examine the durations of LoS and NLoS conditions, and find that *both LoS and NLoS durations follow a power law distribution*, which implies that not only the probability of meeting long LoS conditions is high but also the probability of seeing long NLoS conditions is high. We further investigate the interactions between LoS and NLoS conditions by examining the distribution of PIR and packet inter-loss (PIL) times, which refers to the interval of time elapsed between two dropped packets. We have two key insights as follows. First, *PIR time follows an exponential distribution in LoS conditions but a power law in NLoS conditions.* This means that consecutive packet reception failures can rarely appear when in LoS conditions but can constantly appear when in NLoS conditions. This is cross verified by the observation that PIL time follows a power law distribution in LoS conditions but an exponential in NLoS conditions. Second, unlike the observation that PIR time follows a power law distribution reported in work [22], *the overall PIR time distribution is actually a mix of an exponential distribution of small PIR times in LoS conditions and a power law distribution of PIR times in NLoS conditions.* With those insights, we then discuss three link-aware V2V communication paradigm designs: (1) reliable road-safety message broadcasting; (2) efficient routing establishment; and (3) smart medium resource allocation. The main contributions are summarized as follows.

- We implement an 802.11p-based V2V communication testbed and collect large-volume beaconing traces under three different urban scenarios. In addition, we have labeled LoS and NLoS conditions by watching the recorded videos, and opened the data for public access.[3]

- In general, "perfect zone" in terms of PDR is prevalent over all urban environments. Nevertheless, NLoS conditions induced by large blocking vehicles or slopes can cause severe link performance degradation. In addition, the impact of signal power attenuation to the link performance is not obvious at least within a sufficiently long range of 500 m.
- Frequent packet loss can be found in NLoS conditions, and the distribution of PIL times follows an exponential distribution whereas that of the PIR times follows a power low. Moreover, severe NLoS conditions can last for rather long periods of time. In contrast, link performance is rather reliable in LoS conditions despite a long distance between two communicating vehicles. In LoS conditions, the distribution of PIR times follows an exponential distribution whereas that of the PIL times follows a power law.
- With leveraging the found insights, we organize a discussion on link-aware V2V communication protocol/scheme design to enhance the communication performance.

The remainder of this chapter is organized as follows. Section 5.2 describes the experiment platform and data-collection campaigns. We check the overall performance of 802.11p and delve into the key factor of link performance degradation in Sect. 5.3. In Sect. 5.4, we further investigate the interaction of LoS and NLoS channel conditions and their impacts on 802.11p performance. We organize a discussion on link-aware communication paradigm design in Sect. 5.5. Section 5.6 gives a brief summary.

5.2   Collecting V2V Trace

5.2.1   Experiment Platform Description

In this section, we introduce our V2V communication testbed and the data collection campaign. As shown in Fig. 5.1, the testbed includes two experimental vehicles, each equipped with the following components:

DSRC Module   The off-the-shelf Arada LocoMateTM OBU [23–25] is adopted as the DSRC module, which is mounted on the roof of the experimental vehicle. In the DSRC module, IEEE 802.11p and IEEE 1609 standards are implemented for wireless access in vehicular environments (WAVE). Figure 5.2 shows the WAVE protocol stack (the gray blocks are not involved), where IEEE 802.11p serves as the physical and MAC layer to cope with fast fading and Doppler frequency shift.

---

[3]The labeled trace can be downloaded from ''http://lion.sjtu.edu.cn/project/projectDetail?id=14''.

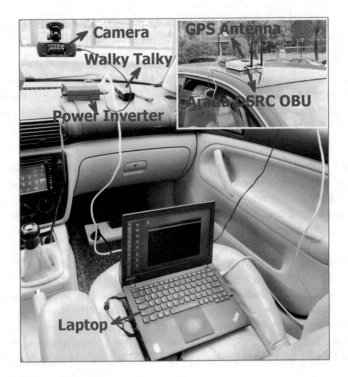

Fig. 5.1 Illustration of an experiment car

The DSRC radio[4] operates in the frequency ranging from 5.700 to 5.925 GHz, and supports one Control Channel (CCH) and multiple Service Channels (SCHs) with two optional bandwidths of 10 and 20 MHz.

For the 10 and 20 MHz channels, the supported data rates range from 3 to 27 Mbps, and 6 to 54 Mbps, respectively. The transmission power can be dynamically specified with the maximum value up to 14 dBm. In our experiments, we adopt the 10 MHz channel with the lowest data rate of 3 Mbps and the maximum transmission power of 14 dBm in order to achieve the most reliable V2V communication (i.e., the derived results thus provide the "upper bound" of performance of 10 MHz 802.11p channels). In addition, the DSRC module has one 680 MHz MIPS processor running Linux, a 16 MB Flash, a 64 MB memory and 1 GB Ethernet interface.

GPS Module A high-performance GPS receiver is integrated in each OBU with an external RF antenna. The GPS receiver can receive the location information

---

[4]Note that, we adopt the single-radio system since the commodity DSRC radio currently cannot support the MIMO system, and the channel switching scheme is disabled as we investigate the V2V communication capability rather than the multi-channel operations.

(including the latitude, longitude, altitude, velocity, etc.) of the experimental vehicle. Besides, the GPS modules are utilized to synchronise both OBUs every 200 ms.

Mobile Computer We use a ThinkPad X240 laptop, to connect and control the OBU via its Gigabit Ethernet interface by running the `telnet` protocol. In addition, as the storage and memory of the OBU are very limited, we buffer all transmitted and received packets at the OBU temporarily, and periodically download those packets to the laptop, in order to keep the data collection program running all the time.

Camera Recorders As the urban environments are highly dynamic and complex, we deploy two cameras on each vehicle with one mounted on the front window and the other fixed on the rear window, in order to record the whole data collection process for offline analysis. The time of cameras are synchronized to the OBU within a precision of one second level.

V2V Beaconing Application By adopting the Wave Short Message Protocol (WSMP), we implement a beaconing application on the WAVE protocol stack which is shown in Fig. 5.2. WSMP is a transport layer protocol, in which there is no retransmission or ACK mechanisms (similar to UDP). There are two programs, i.e., one *transmitter* and one *receiver*, in our beaconing application. Specifically, the receiver keeps listening to the channel and the transmitter transmits a 300-byte beacon (the maximum payload of a WSMP packet can support 1300 bytes)

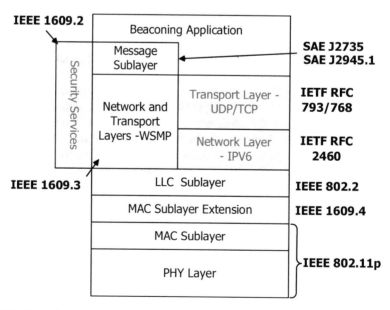

Fig. 5.2 Beaconing application implemented on WAVE protocol stack, where grey blocks are not involved

every 100 ms in accordance with the road safety requirement. Each beacon contains a sequence number as well as the latitude, longitude, altitude, and speed information of the transmitter. Both the transmitter and receiver log the beacon transmission/reception record. By offline comparing the transmitted packets and the received packets, the performance of 802.11p V2V communication can be evaluated.

5.2.2   Data Collection Campaign

To cover all typical urban road conditions, we consider three major road types: (1) *urban*: roads can be unidirectional 1- or 2-lane wide and bidirectional 4- to 8-lane wide, with a large number of tunnels, overhead bridges, tall buildings, and elevated roads, as well as heavy traffic; (2) *suburban*: roads are normally bidirectional and 4- to 6-lane wide with open lands, remote houses, and light traffic; (3) *highway*: bidirectional 8-lane urban freeway with a large number of walls and time-varying traffic.

We conduct our data collection campaigns within areas of the above three road types in Shanghai, and the collection areas are shown in Fig. 5.3. For each road type, the data collection lasts for at least 10 days, and in each day collection, we conduct data collection during two different time periods, i.e., rush hour (from 5:00 pm) and off-peak time (from 8:00 pm). To guarantee valid communication, during experiments, we control the distance between two communicating vehicles to be

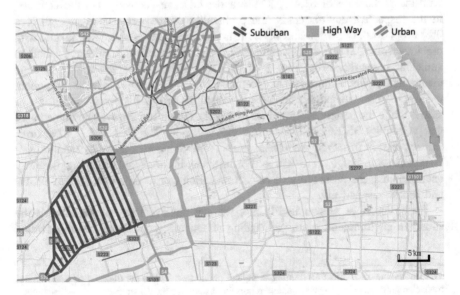

Fig. 5.3 Various urban environments are selected to conduct data collection

no more than 500 m.[5] To mimic realistic driving conditions, there is no additional requirement on how the drivers drive. The overall campaign lasts for more than 2 months with an accumulated distance of over 1500 km. As a result, for each road condition, we obtain a trace, denoted by **trace** $\mathcal{U}$ (urban condition), **trace** $\mathcal{S}$ (suburban condition) and **trace** $\mathcal{H}$ (highway condition). The total amount of all traces adds up to 110 GB. In addition, we concatenate all three traces of different environments together to form a universal trace, denoted by **trace** $\mathcal{A}$.

5.3 Overall Urban V2V Performance Analysis

To gain an overall picture of V2V communication, we first examine the PDR performance in different urban environments. In practice, the PDR is often calculated as the ratio of the number of data packets received at the receiver to the total number of packets transmitted at the transmitter within a pre-defined time window.[6]

5.3.1 Observing Prevalent Perfect Zone

Figure 5.4 shows the cumulative distribution functions (CDFs) of PDR for all traces and it can be seen that the ideal case of V2V communication could frequently happen. Ideal case means all packets are successfully received (i.e., PDR = 100%), and in urban, suburban and highway environments, it happens with the probability of 81.4%, 92.9% and 67.8%, respectively. On the contrary, in the respective environments, the probability of the worst case (i.e., PDR = 0%) drops to 4.3%, 0.7% and 6.9%. It is interesting to compare our results with previous work [21] that studies communication characteristics in rural and suburban vehicular networks. From Fig. 2 of [21], the authors observed the "gray-zone phenomenon" where intermediate reception ($20\%{\leq}PDR{\leq}80\%$) prevails throughout the whole communication range. The probability of this happening reaches over 50.6% while the perfect reception ($PDR{\geq}80\%$) zone is not always guaranteed with the probability 35.2%. *Unlike their observation, we find that 802.11p performs rather reliably in urban environments and "perfect zone" prevails with a wide communication range up to 350 m.* For instance, in urban, suburban and highway environments, the probability of perfect reception can reach above 89.6%, 95.4% and 76.2%, respectively.

Furthermore, as shown in Fig. 5.4, we can observe that compared with the suburban environment, multi-path fading effects are much more severe in urban

---

[5]We implement an application running in mobile phone to calculate the distance of two vehicles by exchanging the GPS information, and report the distance to the driver every 3 s.

[6]In this chapter, we calculate PDR with a time window of 1 s.

Fig. 5.4 CDFs of PDR over all traces

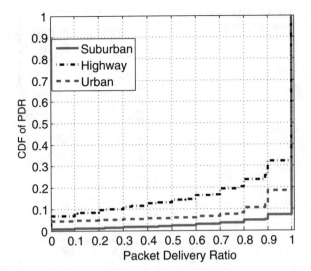

and highway environments. Particularly, in the urban and highway environment, the probability of poor reception (PDR$\leq$20%) is about 5% and 9.6%, respectively, while in the suburban environment, the probability falls to 1.2%. It is reasonable as in the suburban environment, there are few vehicles or obstacles that could cause multipath effects. On the contrary, there are a large number of mobile scatters (high-speed vehicles) and numerous stationary scatters (buildings) in the urban and highway environments, which could inject multiple paths into the channel, resulting in poor PDRs in both environments.

5.3.2 Analyzing Key Factors of Performance Degradation

To derive the key factor of performance degradation, the impact of the communication distance is then investigated. We plot the average PDR within different distance ranges, which is shown in Fig. 5.5. It can be seen that in all studied environments, with the distance increasing, the average PDR drops gradually. However, it is surprising to find that the PDR variation increases dramatically as the communication distance increases, especially for the urban environment. Particularly, supposing at a communication distance of 400 m in the urban environment, the average PDR can often reach up to 100% but can also fall to below 10%. To figure out the reason for such large PDR variations, we check with the recorded videos and observe that packets are frequently lost when two vehicles are blocked by obstacles, i.e., encountering NLoS conditions. To this end, based on watching videos, we then

Fig. 5.5 PDR vs. distance between vehicles

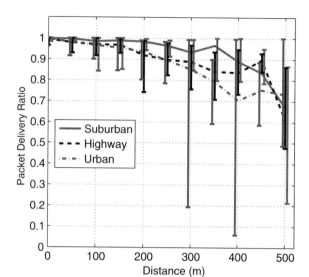

mark all NLoS situations when two vehicles cannot visually see each other,[7] and divide the original trace into LoS and NLoS two categories. In real driving scenarios, between two communicating vehicles, there may be slopes, big obstacles (e.g., trucks and buses), and turns which could result in NLoS conditions.

NLoS Conditions Instead of Separation Distance Affect Link Performance Most Figure 5.6 shows the CDFs of PDR in LoS and NLoS conditions, respectively, and we can see that most packet reception failures happen under NLoS conditions. For example, in the urban, suburban and highway environments, the probability of poor reception (PDR≤20%) under NLoS conditions reaches over 82.6%, 48.3% and 62.1%, respectively, and the probability of perfect reception is zero in all environments. On the contrary, in the urban, suburban and highway environments, the probability of perfect reception under LoS conditions is 93.5%, 96.9% and 86%, respectively, and the probability of poor reception is less than 1% in all environments. We can conclude that *it is NLoS conditions instead of separation distance that lead to most packet reception failures*. Although the separation distance is not the direct reason of poor PDR, it is true that the probability of encountering a NLoS condition increases as the separation distance increases, which explains the large PDR variations at long separation distances. The insight can be further verified by Fig. 5.7, which shows results of Fig. 5.5 in LoS and NLoS

---

[7]Note that, although NLoS conditions found by cameras are not necessarily to be NLoS for RF radios, those visually NLoS conditions are still good approximations of real radio NLoS conditions and valuable for analysis.

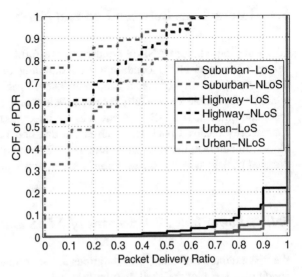

Fig. 5.6 CDFs of PDR in LoS/NLoS conditions

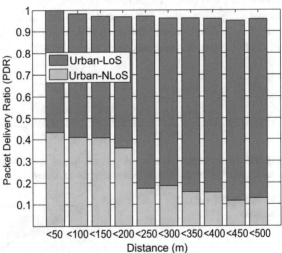

Fig. 5.7 PDR vs. distance under LoS/NLoS conditions

conditions, respectively.[8] We can see that the average PDRs in LoS conditions seem to be rather stable (all above 95%) while the average PDRs in NLoS conditions have poor performance (all below 40%) regardless of the distance variation.

---

[8]Note that, results under suburban and highway scenarios are omitted due to the similar observation and space limitation.

5.4   Interactions Between LoS and NLoS

Given the importance of NLoS conditions, we further investigate the interactions between LoS and NLoS conditions by examining the metrics of PIR and PIL times. As shown in Fig. 5.8, a time sequence of packets are present, where the white blocks denote successfully received packets and dark blocks represent packet reception failures. The PIR time refers to the interval of time elapsed between two successfully received packets which is the duration between two adjacent white blocks. In contrast, the PIL time refers to the interval of time elapsed between two dropped packets which is the duration between two adjacent dark blocks.

5.4.1   Power Law Distributions of NLoS and LoS Durations

We first examine LoS and NLoS durations to check how they appear in real driving conditions, and plot their tail distributions, which are shown in Figs. 5.9 and 5.10, respectively. Two main observations can be achieved. First, *both LoS and NLoS durations follow a power law distribution*, as linear plots in log-log scale are found in both figures. It indicates that not only the probability of meeting long LoS

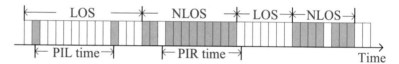

Fig. 5.8 An example sequence of packets, where white blocks denote successfully received packets and dark blocks denote packet reception failures

Fig. 5.9 CCDFs of LoS durations

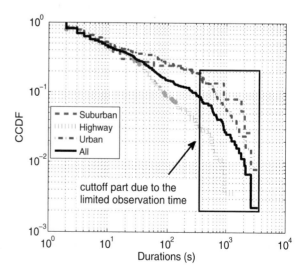

Fig. 5.10 CCDFs of NLoS durations

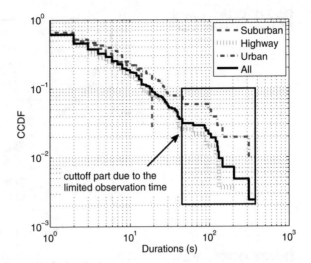

conditions is high but also the probability of meeting long NLoS conditions is also high. It should be noted that, the cutoff part of the tail distribution should not be considered due to the effect of limited observation duration, which has also been pointed out in previous studies on characterizing inter-contact time distribution of human [26] and vehicular [27] mobility. Second, LoS durations are in general longer than NLoS durations. For instance, the proportion of durations longer than 10 s reaches 50% in LoS conditions while drops to only 18% in NLoS conditions. Nevertheless, heavy-tailed NLoS durations have important implications on the design of beacon-based road-safety applications, which shall be able to cope with relatively long and constant communication blackouts.

5.4.2   Mixed Distributions of PIR Times

Considering the importance of NLoS conditions, we further investigate the interactions between LoS and NLoS conditions by examining the distribution of PIR and PIL times. Figure 5.11 shows the complementary cumulative distribution functions (CCDFs) of PIR times of all traces in log-log scale. It is also interesting to compare our results with previous work [22] that studied 802.11p-based beaconing performance based on data collected during trips traveled among several Italian cities. As shown in Fig. 4 of [22], the authors observed that the CCDF of PIR times satisfies a power law (identified by linear plots in log-log scale) and had the conclusion that the PIR time distribution is heavy tailed, which means that the probability of having relatively long PIR time is relatively high. *Unlike their observation, we find that the CCDF of PIR time appears linear in log-log scale only for large PIR times and has a much faster decay for small PIR times, which*

Fig. 5.11 CCDFs of overall
PIR times, in log-log scale

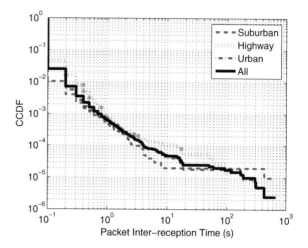

Fig. 5.12 CCDFs of PIR
times in LoS conditions, in
linear-log scale

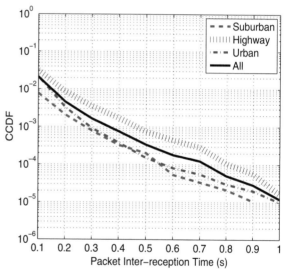

implies that the PIR time only partially follows a power law. For example, as shown
in Fig. 5.11, it can be seen that the CCDF of PIR times is not linear when PIR time
is smaller than 1 s.

To explain it, we can see from Fig. 5.6, where large proportion of PDRs are
greater than 80% in LoS conditions while in NLoS conditions, very poor PDR
is witnessed, indicating that small PIR times are common in LoS conditions and
large PIR times normally appear in NLoS conditions. To this end, we check the
distribution of PIR times in LoS and NLoS conditions, and plot their CCDF results
in Fig. 5.12 (linear-log scale) and Fig. 5.13 (log-log scale), respectively. Clear linear
plots are seen in both figures, which means that *PIR time in LoS conditions follows
an exponential distribution, but follows a power law distribution in NLoS condi-*

Fig. 5.13 CCDFs of PIR times in NLoS conditions, in log-log scale

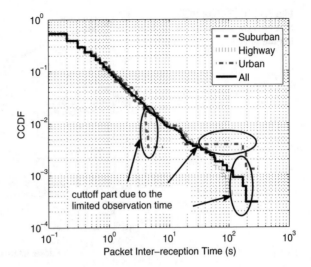

tions. It implies that short PIR times (consecutive successfully-received packets) are more likely to happen in LoS conditions whereas the probability of having long PIR time is relatively high under NLoS conditions. Then, we can well explain why the CCDF of overall PIR times in Fig. 5.11 has a much faster decay than a power law distribution when the PIR times are small. It is because the resulted CCDF is a combination of an exponential distribution of small PIR times in LoS conditions, a power law distribution of PIR times in NLoS conditions and a power law distribution of NLoS durations. The main reason that we have distinct observations from the previous work [22] where the overall distribution of PIR times follows a power law, may be because the authors did not discriminate between LoS and NLoS conditions in their analysis, and NLoS conditions might take an inappropriately large portion during the field testing. On the other hand, we have a similar but opposite observation on the distribution of PIL times. In particular, as shown in Figs. 5.14 and 5.15, we find that *PIL time in LoS conditions follows a power law distribution whereas that in NLoS conditions follows an exponential distribution.* It means that short PIL times (consecutive packet losses) are common in NLoS conditions while in LoS conditions, relatively long PIL times are more likely to happen. This is reasonable as frequent packet failures are more likely to happen in NLoS conditions.

5.4.3 Severe NLoS Condition Hurts

According to our previous analytics, we have confirmed that NLoS conditions can last relatively long with poor PDRs. To further understand how the DSRC performs under different NLoS conditions, i.e., *severe* (PDR≤20%), *intermediate*

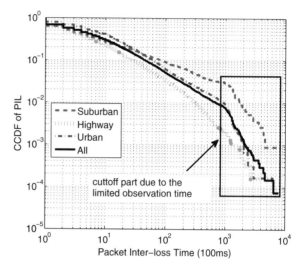

Fig. 5.14 CCDFs of PIL times in LoS conditions, in log-log scale

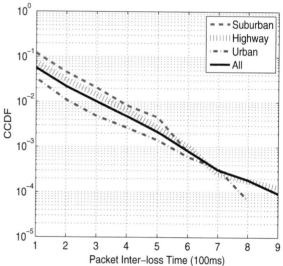

Fig. 5.15 CCDFs of PIL times in NLoS conditions, in linear-log scale

(20%<PDR≤40%), and *normal* (40%<PDR≤70%), the distributions of NLoS durations and PIL times are examined by using the combined **trace** $\mathcal{A}$. As shown in Fig. 5.16, we can observe that severe NLoS conditions usually results in relatively longer durations. For instance, 20% durations of severe NLoS conditions are longer than 1 min, but the duration of intermediate and normal NLoS conditions are normally very short. Additionally, the longest duration under intermediate and normal NLoS conditions is about 40 and 30 s, respectively, but the value can reach up to 380 s when under severe NLoS conditions. It means that when meeting severe NLoS conditions, the system would suffer from relatively long and constant communication blackouts, which should be carefully coped with. On the other hand,

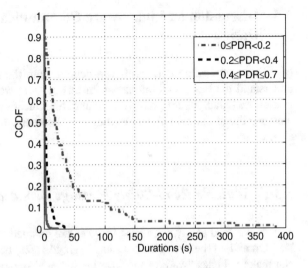

Fig. 5.16 CCDF of durations of different NLoS types

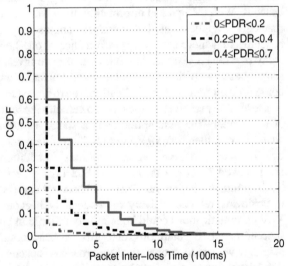

Fig. 5.17 CCDF of PIL times in different NLoS conditions

as shown in Fig. 5.17, when under severe NLoS conditions, we can observe that the probability of a PIL time ≤100 ms (i.e., equaling the beacon interval) reaches above 95%. It indicates that if the previous packet is lost, the current packet will also be lost with a high probability, i.e., ≥95%. However, when under intermediate and normal NLoS conditions, the probability drops down to 70% and 40%, respectively. This inspires us that necessary retransmissions are required especially under severe NLoS conditions, since the communication link is seriously intermittent.

5.5 Discussion on Link-Aware Communication Paradigm Design

Many communication paradigms can benefit from the link knowledge if we can understand the link condition ahead. In this section, we organize a discussion on link-aware paradigm design for vehicular communications, i.e., reliable road-safety message broadcasting, efficient routing establishment, and smart medium resource allocation

5.5.1 Reliable Road-Safety Message Broadcasting

To achieve road safety, broadcast services are enabled in the IEEE 802.11p-based DSRC standard (i.e., vehicles broadcasting road-safety beacons periodically), which can make vehicles aware of the surrounding environment by receiving up-to-date beacons. With the updated information, the system can support various upper-layer road-safety applications, but potential dangers may arise when with any loss of beacons. To achieve satisfying performance of one-hop broadcast communications, many previous researches in the literature have investigated how to transmit road-safety beacons with guaranteeing their transmission requirements. However, a common assumption is made in all these works, where the transmitted beacon can be successfully received as long as two vehicles are within the communication range of each other. The assumption cannot stand according to our field experiments, where it is verified that V2V links are intermittent and can be significantly affected by driving contexts, e.g., NLoS conditions. In addition, as RTS/CTS/ACK packets are removed in broadcast mode of 802.11p to facilitate response, vehicles can hardly confirm whether the transmitted beacons are well received, posing challenges in achieving reliable delivery of road-safety beacons. Based on the insight that V2V communications could be relatively reliable in LoS conditions but can hardly succeed in NLoS conditions, one possible beaconing strategy is to find a proper neighbor vehicle to cooperatively rebroadcast beacons when meeting harsh NLoS conditions. For example, as shown in Fig. 5.18a, if the vehicle A and vehicle B are blocked by obstacles where the vehicle B can hardly receive the beacons from the vehicle A, to enhance the beaconing reliability, the vehicle A can ask the vehicle D to rebroadcast the beacon. We refer to the vehicle D as a *helper* vehicle. To optimize the system performance, for all optional helpers, the helper with the best link quality with both the sender and receiver should be selected. Under this rule, the system would not choose the vehicle C since it has an NLoS link condition with the sender. This link-aware beaconing strategy can improve the reliability of beacon delivery without costing much rebroadcast overhead, since only required rebroadcasts are triggered with efficient retransmission. However, to achieve such paradigm, the system shall be able to precisely identify link conditions and choose

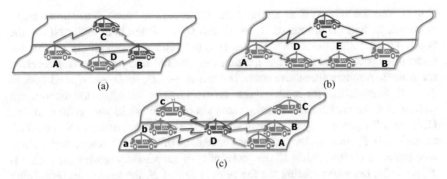

Fig. 5.18 Link-aware communication paradigms design. (**a**) Broadcasting beacons reliably. (**b**) Efficient routing establishment. (**c**) Smart medium resource allocation

suitable helpers efficiently in real time, both which need further in-depth research, and we leave them as our focuses in the next chapter to enhance the link reliability.

5.5.2   Efficient Routing Establishment

For two remote vehicles in VANETs, data packets between them can be exchanged via multiple-hop transmissions. Vehicles could communicate with each other without requiring the presence of infrastructure but based on cooperation. Particularly, intermediate vehicles can act as routers to forward packets to the desired vehicle who is out of the radio coverage of sending vehicle. However, with the intermittent wireless link and network topology varying fast, routing in vehicular networks becomes technically challenging. On the one hand, if the system maintains a routing table to guide the route discovery, the source vehicle can look up the table when initiating a routing process. However, as vehicles move fast, the table information cannot be accurately updated with guaranteeing the correctness of routing path. As a result, severe packet loss could happen with either the error routes or unreliable wireless link conditions. On the other hand, if the system adopts the broadcasting to discover correct routes and vehicles blindly retransmit the received packets (i.e., flooding), the channel may witness an explosive growth of traffic with severe channel congestion (i.e., broadcast storm problem). Consequently, frequent transmission collisions could happen, where the transmissions can hardly succeed. To improve the routing performance, it is promising to take link conditions into consideration, i.e., routing packets to those links with good qualities. For instance, as shown in Fig. 5.18b, when the vehicle A wants to transmit packets to the vehicle B who is out of the direct radio coverage, the routing process is required. Under this case, the vehicle D and E is within the one-hop range of the sender A and receiver B, respectively, and the vehicle C is within the one-hop range of both vehicle A and B. To route a packet from the vehicle A to B, one plausible approach is to choose the vehicle C as a relay since the route can have the minimum hops. However, as link conditions of $A \to C$ and $C \to B$ are both under NLoS, the two links are rather unreliable. To this end, the vehicle C can hardly receive the transmitted packet and the retransmission can also fail likely even having received the packet. Another alternative route is from $A \to D$, to $D \to E$, and then to $E \to B$. Even though the route requires an extra transmission hop, the receiver can well receive the packet since it is delivered via three LoS links with good qualities. Given that the probability of successfully receiving a packet under LoS and NLoS conditions is P_L and P_N, they can be assumed to be 0.95 and 0.2, respectively.[9] Then, when routing via the vehicle C, the probability of successfully routing the packet is $P_N^2 = 0.04$, but when routing via the vehicle D and E, the successful probability could reach $P_L^3 = 0.86$. To this end, we can conclude that the route performance can be significantly affected by the link conditions, which should be considered as a driving factor in casting efficient routing strategies.

5.5.3 Smart Medium Resource Allocation

In VANETs, the MAC protocol is of significant importance for upper-layer applications, and different applications generally have distinct medium access demands. Specifically, for road-safety applications (e.g., collision avoidance and pre-crash sensing), ultra-low delay (normally within 100 ms) is required, which means that vacant medium resource should be reserved whenever to feed road-safety applications. In contrast, comfort applications (e.g., video streaming and file downloading) can somewhat tolerate a certain delay while requiring substantial spectrum resources. However, only 75 MHz, 5.9 GHz licensed spectrum is allocated to DSRC by the FCC, which is insufficient to deliver quality services for high-density vehicles. To this end, the medium resource allocation becomes a significant role to affect the performance of application provisioning. In current VANET MAC protocols like contention-based MAC 802.11p [28] and TDMA-based MACs [29, 30], the system treats all vehicles equally in terms of the channel access, where link conditions are not considered. With scarce medium resources, the equal-fairness scheme may degrade the system performance, especially under high-density scenarios, since when meeting NLoS conditions, transmissions can hardly succeed but incur interferences to neighboring vehicles. To improve the resource utilization, one smarter strategy is to allocate the medium resources to those vehicles having good-quality links with receivers [31]. For instance, as shown in Fig. 5.18c, when each vehicle wants to access the channel and transmit data to the vehicle D, the equal-fairness manner of medium allocation may lose efficiency considering the obstacles between communicating vehicles. Particularly, if there are big obstacles such as buses or trucks following the vehicle D, the communication links between

---

[9] The values are empirically set based on the average PDRs under LoS and NLoS conditions.

the vehicle D and the following vehicles a, b and c likely have NLoS conditions. On the contrary, the communication links between the vehicle D and the heading vehicles A, B and C are with LoS conditions. Obviously, compared with the vehicles a, b and c, the vehicles A, B and C should have a higher priority to access the channel. In 802.11p, based on the CSMA/CA mechanism, vehicles negotiate the channel usage with the distributed coordination function. To be specific, the vehicle has to sense the channel before accessing the medium; the vehicle can access the channel if it is sensed to be free, otherwise the vehicle has to perform the random back-off procedure. Under this case, the system can determine the back-off time unequally for vehicles with considering their link qualities. In doing so, when under light-traffic scenarios, vehicles with good-link conditions can access the channel with smaller delays, and the resource utilization can be further enhanced when under heavy-traffic scenarios. On the other hand, in TDMA-based MACs, time is partitioned into frames, each containing a fix number of equal-length time slots. During each time slot, each vehicle decides whether to transmit a packet or not with a probability p. To achieve a fair time slot acquisition, the system can dynamically determine the value p for vehicles in accordance with their link conditions. In the broadcast paradigm for road-safety applications, the system can leverage the information of the number of LoS links $|LoS|$ and NLoS links $|NLoS|$ of vehicles that they have with their one-hop neighbors, as an effective indicator for medium resource allocation.

5.6 Summary

In this chapter, we have studied the 802.11p-based V2V communications in urban environments based on real-world data traces. We have obtained the following two major insights. First, 802.11p works very reliably in urban settings with a wide range of "perfect zone" found. Second, LoS and NLoS channel conditions are crucial for reliable V2V communications. In particular, they have very opposite characteristics with respect to the PIR and PIL time distributions. The intervals between a pair of successfully received packets have an exponential distribution in LoS conditions but turns out to be a power law when in NLoS conditions. In addition, we have organized a discussion on how to utilize the unique characteristics of V2V link behaviors in order to improve the performance of vehicular networking applications.

References

1. W. Zhuang, Q. Ye, F. Lyu, N. Cheng, J. Ren, SDN/NFV-empowered future IoV with enhanced communication, computing, and caching. Proc. IEEE **108**(2), 274–291 (2020)
2. ASTM, Standard Specification for Telecommunications and Information Exchange Between Roadside and Vehicle Systems - 5 GHz Band Dedicated Short Range Communications (DSRC) Medium Access Control (MAC) and Physical Layer (PHY) Specifications. http://www.astm.org/Standards/E2213.htm

3. F. Lyu, N. Cheng, H. Zhu, H. Zhou, W. Xu, M. Li, X. Shen, Towards rear-end collision avoidance: adaptive beaconing for connected vehicles. IEEE Trans. Intell. Transp. Syst. 1–16, Early Access (2020). https://doi.org/10.1109/TITS.2020.2966586
4. H. Peng, D. Li, K. Abboud, H. Zhou, H. Zhao, W. Zhuang, X. Shen, Performance analysis of IEEE 802.11p DCF for multiplatooning communications with autonomous vehicles. IEEE Trans. Veh. Technol. **66**(3), 2485–2498 (2017)
5. M.A. Togou, L. Khoukhi, A. Hafid, Performance analysis and enhancement of WAVE for V2V non-safety applications. IEEE Trans. Intell. Transp. Syst. **19**(8), 2603–2614 (2018)
6. W. Xu, H. Zhou, N. Cheng, F. Lyu, W. Shi, J. Chen, X. Shen, Internet of vehicles in big data era. IEEE/CAA J. Autom. Sin. **5**(1), 19–35 (2018)
7. S. Zhang, J. Chen, F. Lyu, N. Cheng, W. Shi, X. Shen, Vehicular communication networks in the automated driving era. IEEE Commun. Mag. **56**(9), 26–32 (2018)
8. F. Abbas, P. Fan, Z. Khan, A novel low-latency V2V resource allocation scheme based on cellular V2X communications. IEEE Trans. Intell. Transp. Syst. Early Access (2018). https://doi.org/10.1109/TITS.2018.2865173
9. F. Lyu, H. Zhu, H. Zhou, L. Qian, W. Xu, M. Li, X. Shen, MoMAC: mobility-aware and collision-avoidance MAC for safety applications in VANETs. IEEE Trans. Veh. Technol. **67**(11), 10590–10602 (2018)
10. X. Zheng, W. Chen, P. Wang, D. Shen, S. Chen, X. Wang, Q. Zhang, L. Yang, Big data for social transportation. IEEE Trans. Intell. Transp. Syst. **17**(3), 620–630 (2016)
11. H. Zhou, N. Cheng, Q. Yu, X. Shen, D. Shan, F. Bai, Toward multi-radio vehicular data piping for dynamic DSRC/TVWS spectrum sharing. IEEE J. Sel. Areas Commun. **34**(10), 2575–2588 (2016)
12. N. Cheng, F. Lyu, J. Chen, W. Xu, H. Zhou, S. Zhang, X. Shen, Big data driven vehicular networks. IEEE Netw. **32**(6), 160–167 (2018)
13. X. Cheng, C. Chen, W. Zhang, Y. Yang, 5G-enabled cooperative intelligent vehicular (5Gen-CIV) framework: when Benz meets Marconi. IEEE Intell. Syst. **32**(3), 53–59 (2017)
14. J. Liang, Z. Qin, S. Xiao, L. Ou, X. Lin, Efficient and secure decision tree classification for cloud-assisted online diagnosis services. IEEE Trans. Dependable Secure Comput. 1–13 (2019). https://doi.org/10.1109/TDSC.2019.2922958
15. R. Meireles, M. Boban, P. Steenkiste, O. Tonguz, J. Barros, Experimental study on the impact of vehicular obstructions in VANETs, in *Proceedings of IEEE VNC*, Dec 2010
16. M. Boban, J. Barros, O. Tonguz, Geometry-based vehicle-to-vehicle channel modeling for large-scale simulation. IEEE Trans. Veh. Technol. **63**(9), 4146–4164 (2014)
17. I. Tan, W. Tang, K. Laberteaux, A. Bahai, Measurement and analysis of wireless channel impairments in DSRC vehicular communications, in *Proceedings of IEEE ICC*, May 2008
18. L. Cheng, B. Henty, D. Stancil, F. Bai, P. Mudalige, Mobile vehicle-to-vehicle narrow-band channel measurement and characterization of the 5.9 GHz dedicated short range communication (DSRC) frequency band. IEEE J. Sel. Areas Commun. **25**(8), 1501–1516 (2007)
19. X. Yin, X. Ma, K. Trivedi, A. Vinel, Performance and reliability evaluation of BSM broadcasting in DSRC with multi-channel schemes. IEEE Trans. Comput. **63**(12), 3101–3113 (2014)
20. M. Soltani, M. Alimadadi, Y. Seyedi, H. Amindavar, Modeling of Doppler spectrum in V2V urban canyon oncoming environment, in *Proceedings of 7th IST*, Jan 2014
21. F. Bai, D.D. Stancil, H. Krishnan, Toward understanding characteristics of dedicated short range communications (DSRC) from a perspective of vehicular network engineers, in *Proceedings of ACM MobiCom*, Sept 2010
22. F. Martelli, M. Elena Renda, G. Resta, P. Santi, A Measurement-based study of beaconing performance in IEEE 802.11p vehicular networks, in *Proceedings of IEEE INFOCOM*, May 2012
23. H. Zhou, W. Xu, J. Chen, W. Wang, Evolutionary V2X technologies toward the internet of vehicles: challenges and opportunities. Proc. IEEE **108**(2), 308–323 (2020)
24. Arada Systems, LocoMate classic on board unit OBU-200. http://www.aradasystems.com/locomate-obu/

25. F. Lv, H. Zhu, H. Xue, Y. Zhu, S. Chang, M. Dong, M. Li, An empirical study on urban IEEE 802.11p vehicle-to-vehicle communication, in *Proceedings of IEEE SECON*, June 2016
26. A. Chaintreau, P. Hui, J. Crowcroft, C. Diot, R. Gass, J. Scott, Impact of human mobility on opportunistic forwarding algorithms. IEEE Trans. Mobile Comput. **6**(6), 606–620 (2007)
27. H. Zhu, L. Fu, G. Xue, Y. Zhu, M. Li, L. Ni, Recognizing exponential inter-contact time in VANETs, in *Proceedings of IEEE INFOCOM*, March 2010
28. C. Han, M. Dianati, R. Tafazolli, R. Kernchen, X. Shen, Analytical study of the IEEE 802.11p MAC sublayer in vehicular networks. IEEE Trans. Intell. Transp. Syst. **13**(2), 873–886 (2012)
29. F. Lyu, H. Zhu, H. Zhou, W. Xu, N. Zhang, M. Li, X. Shen, SS-MAC: a novel time slot-sharing MAC for safety messages broadcasting in VANETs. IEEE Trans. Veh. Technol. **67**(4), 3586–3597 (2018)
30. H.A. Omar, W. Zhuang, L. Li, VeMAC: a TDMA-based MAC protocol for reliable broadcast in VANETs. IEEE Trans. Mobile Comput. **12**(9), 1724–1736 (2013)
31. F. Lyu, N. Cheng, H. Zhou, W. Xu, W. Shi, J. Chen, M. Li, DBCC: leveraging link perception for distributed beacon congestion control in VANETs. IEEE Internet Things J. **5**(6), 4237–4249 (2018)

Chapter 6
Link-Aware Reliable Beaconing Scheme Design

After characterizing the link communication performance, we investigate the link-aware beaconing scheme design in this chapter. Particularly, we confirmed that among many types of contextual information, NLoS link condition is the key factor of V2V performance degradation, which can significantly deteriorate the beaconing reliability. In this chapter, we propose a link-aware reliable beaconing scheme, named *CoBe* (i.e., *Co*operative *Be*aconing), to enhance the broadcast reliability for road-safety applications. *CoBe* is a fully distributed scheme, in which a vehicle first detects the link condition with each neighbor by machine learning algorithms, then exchanges such link condition information with its neighbors, and finally selects the minimal number of helper vehicles to rebroadcast its beacons to those neighbors in bad link conditions. To analyze and evaluate the performance of *CoBe* theoretically, we devise a two-state Markov chain model to mimic beaconing behaviors under LoS/NLoS conditions. In addition to *CoBe*, we also present a case study of efficient unicasting scheme for non-road-safety applications. Extensive trace-driven experiments are carried out, and the results demonstrate the efficacy of both broadcasting and unicasting schemes.

6.1 Problem Statement

Considering the significant impact of NLoS conditions on the link performance, i.e., massive blackout events are more likely to occur under NLoS conditions, the performance of road-safety application can be greatly affected since they rely on reliable beacon exchanges [1–5]. To this end, in this chapter, we propose a link-aware reliable beaconing strategy, named *CoBe* (i.e., *Co*operative *Be*aconing), to enhance the broadcast reliability when meeting harsh NLoS conditions. In *CoBe*, the link states (LoS or NLoS) among neighbors are first detected in real time by supervised machine learning algorithms. In the payload of each beacon, in addition

© Springer Nature Switzerland AG 2020
F. Lyu et al., *Vehicular Networking for Road Safety*, Wireless Networks,
https://doi.org/10.1007/978-3-030-51229-3_6

to application data, vehicles also include the information of the link states with its one-hop neighbors. Upon identifying an NLoS condition, the sender selects *a helper vehicle* with the best link quality with both the sender and the receiver among all the optional helpers, to rebroadcast its beacons. The LoS condition is not importantly considered since the link quality is rather reliable in LoS conditions (reflected by the exponential distribution of PIR time under LoS conditions), and seeking rebroadcasts can inflict extra overhead over the network. As *CoBe* runs at the application layer and no additional environment input or cross-layer information is required, it is easy and feasible to be implemented in practice. To analyze the performance of *CoBe* theoretically, we devise a two-state Markov chain to model beaconing behaviors with taking LoS/NLoS channel conditions into account, based on which we derive the beaconing reliability and the corresponding cost, evaluated by the metrics of *Beacon Reception Ratio (BRR)* and *Broadcast Utility (BU)*, respectively. Beyond that, by using the power law distributions of LoS and NLoS durations, exponential distribution of PIR time under LoS conditions, and exponential distribution of PIL time under NLoS conditions, we propose a scheme to synthesize V2V link communication data. At last, we conduct extensive trace-driven simulations to evaluate the performance of *CoBe*. Both numerical and experiment results demonstrate the efficacy of *CoBe*, where BRR can be significantly enhanced with a small cost of BU. Additionally, highly analogous performance between the modeling and experiment results verify the accuracy of the proposed Markov chain model, which can be of paramount importance for other VANET researches such as performance analysis, model establishment, and parameter tuning.

In addition to the broadcasting scheme, we also implement a case study of efficient unicast scheme for non-road-safety applications [6, 7]. Particularly, inspired by the intuition that blindly sending more packets in harsh NLoS conditions can hardly succeed but incur resource wasting and increase interferences to other neighbouring vehicles, we propose *Bungee*, to avoid bad opportunities for data transmission. In *Bungee*, once an NLoS condition is identified by a vehicle, the vehicle will adjust its packet sending intervals based on the estimated behaviors of LoS/NLoS conditions to avoid the upcoming bad channel condition, and recover when the channel quality becomes good. We conduct extensive trace-driven experiments to demonstrate the efficiency of *Bungee* in terms of resource utilization enhancement. In general, with efficient resource utilization and less interference, *Bungee* can be adopted directly by delay-tolerated applications, especially under high density scenarios. The main contributions are threefolds.

- We propose a link-aware reliable beaconing strategy, named *CoBe*, which is a fully distributed scheme and integrates three major techniques: (1) online NLoS detection; (2) link status exchange; (3) beaconing with helpers. *CoBe* can play an important role in supporting reliable beacon exchanges for road-safety applications in realistic urban driving scenarios.
- With the field measurement data, we build a two-state Markov chain model to mimic the link performance under LoS/NLoS conditions, which can provide a basis for theoretical analysis. In addition, based on the fitting distributions

of link communication metrics, we devise a scheme to synthesize V2V link communication trace, which can be valuable for other data-driven studies.
• We present a case study of efficient unicast scheme for non-road-safety applications, which can adapt to LoS/NLoS channel conditions, and avoid bad opportunities for data transmission, to enhance the spectrum utilization.

The remainder of this chapter is organized as follows. we elaborates on the design of *CoBe* in Sect. 6.2. We build a two-state Markov chain model to analyze the performance of *CoBe* theoretically in Sect. 6.3. Section 6.4 details the synthesizing process of V2V link communication trace. Performance evaluation is carried out in Sect. 6.5. The case study of efficient unicast scheme is presented in Sect. 6.6. Section 6.7 gives a brief summary.

6.2 Design of CoBe

6.2.1 Overview

Reliable beaconing is an essential building block for road-safety applications, where periodical "status" messages (containing the information of vehicle position, speed, acceleration, braking status, etc.) broadcasted by each vehicle should be successfully received by all neighbors in its vicinity. From our previous analysis, beacon-based road-safety applications could benefit from link context information. For instance, when encountering NLoS conditions, a better beaconing strategy is to find proper neighbor vehicles to help rebroadcast the beacons, to enhance the beaconing reliability. To this end, we propose a link-aware cooperative beaconing strategy, called *CoBe*, to cope with the harsh NLoS conditions. In essence, *CoBe* integrates three major techniques as follows: (1) online NLoS detection; (2) link status exchange; (3) beaconing with helpers.

6.2.2 Online NLoS Detection

Using Physical Layer Hints With more advanced antenna (e.g., MIMO systems) and physical layer techniques (e.g., OFDM modulation), NLoS conditions can be accurately detected in real time. For example, the power delay profile [8], which profiles arriving signals from multi-path channels and gives the power strength of a received signal, can be utilized to detect NLoS conditions.

Utilising Static Topographical Information In addition to dynamic vehicular traffic, many NLoS conditions are caused by slopes and turns between two communicating vehicles. With each vehicle mounted with a GPS receiver, the location information can be easily obtained and shared with each other. By checking with an accurate digital map, situations such as slopes and turns can be accurately identified, which can guide the NLoS identification.

Perceiving NLoS at Application Layer by Machine Learning Algorithms In cases where physical layer hints or topographical information are not available, it is still possible for upper-layer applications to perceive NLoS conditions. From Fig. 5.6, we can observe that all visually identified NLoS conditions have very low PDRs which are less than 70%, while LoS conditions have relatively high PDRs and 95% of them are greater than 70%. This observation indicates that PDR values have some latent relations to the underlying LoS or NLoS channel conditions. Therefore, we can adopt the application layer PDR values to infer the underlying LoS/NLoS conditions. To learn those unknown relations, supervised machine learning algorithms can be well applied since we have marked all LoS and NLoS conditions [9, 10]. Particularly, we can follow the following three steps: (1) Labeling NLoS Conditions: Before using machine learning techniques, we have to first label out all NLoS conditions, and choose one part for model training and the other part for model validation. Since the whole data collection campaigns are recorded by cameras, we label all NLoS conditions when two vehicles cannot visually see each other, and the NLoS and LoS condition is labeled as 0 and 1, respectively; (2) Feature Extraction: To enable efficient machine learning algorithms, it is necessary to extract representative tuple of features from raw data rather than input them directly. We have disclosed that the PDR values can be significantly affected by LoS/NLoS conditions. Therefore, we extract historical PDR values as features for training. Particularly, for each sample, we select three features PDR_1, PDR_5 and PDR_{10}, i.e., the PDR value of previous 1 s, PDR value of previous 5 s, and PDR value of previous 10 s. Then, we can obtain extensive samples in a format of <3-dimensional features, label>; (3) Supervised Machine Learning: After samples constructed, we then use parts of them for model training. Classical machine learning algorithms such as support vector machine (SVM), k-nearest neighbor (KNN), decision tree and random forest, are leveraged to carry out the training process and output a binary classifier. By calculating the previous 1-, 5- and 10-s PDR values and inputting into the classifier, NLoS conditions can be identified in real time. Particularly, if the classifier outputs a result 0 (i.e., representing an NLoS condition), the channel is considered to be in harsh condition; otherwise, an LoS condition is found.

6.2.3 Link Status Exchange

To be compliant with the broadcast requirement for most of road-safety applications, each vehicle is required to broadcast every 100 ms [11–13]. By logging the beacon reception records for each one-hop neighbor, vehicles are able to calculate the up-to-date historical PDR values of each one-hop link, based on which the real-time link status (LoS or NLoS) can be perceived. Once an vehicle identifies a NLoS condition for a link, the vehicle should select a helper vehicle to rebroadcast its beacon, in order to recover the harsh link condition. To assist vehicles in selecting an optimal helper vehicle in a distributed way, in addition to application data, vehicles should also include their one-hop link status information (0 means NLoS and 1 means LoS) in each beacon, in order to exchange such link status among neighbors. Specifically, for each vehicle (say vehicle x), it maintains a circular recording queue for each one-hop link and calculates the up-to-date historical PDRs to get the link status. It then includes the information of (neighbor vehicle ID, LoS/NLoS flag) in each beacon, and broadcasts out. In doing so, with receiving beacons, vehicle x can understand the link status between its one-hop neighbor (say vehicle y) and the neighbors of y, i.e., seeing two-hop links, which benefits to helper selection.

6.2.4 Beaconing with Helpers

Upon identifying one or more NLoS links between a vehicle and its neighbors, the vehicle seeks for helper vehicles from its neighbors to rebroadcast the beacon. Such a helper should be selected if it has an LoS condition with both the sender and the receiver. In general, we define the *helper selection problem* as, in case that there are multiple NLoS links, how to select helpers so that all NLoS links are covered and the number of helpers are minimized at the same time? The following theorem can be concluded.

Theorem 6.1 *The helper selection problem is NP-hard.*

Proof To prove its NP-hardness, we devise a polynomial reduction from a classic NP-hard problem, i.e., *Max k-cover* [14], to the helper selection problem. In the *Max k-cover* problem, there is a collection of subsets, $F = \{S_1, S_2, \ldots, S_i\}$, each being with a set of n points. The objective is to select k subsets from F to maximize the total number of points contained in their union.

The instance of the *Max k-cover* problem can be taken as an input for reduction as follows. In specific, assuming that the link conditions between each pair of vehicles within the communication range of a sender vehicle are known, we can construct a graph, $G(N, E)$, where N is the set of nodes and E is the set of edges. In the graph, each vehicle within the communication range of the sender is a node, and there is an edge between a pair of nodes if they have an LoS link. Denote each node by n_i, and all its LoS neighbors, i.e., the nodes having an LoS link with n_i, by $S_i \subseteq N$, for $i = 1, \ldots, |N|$. With this graph, given the value k, to find k different S_j for $j = 1, \ldots, k$ such that their union contains as many nodes as possible, is equivalent to the *Max k-cover* problem, which is NP-hard. The problem is then to find the smallest number of k to cover all nodes. Therefore, the helper selection problem is an NP-hard problem, which concludes the proof. □

Given the NP-hardness of the helper selection problem, in *CoBe*, we adopt a greedy heuristic to select preferable helpers. In specific, neighbors are ranked according to the size of S_i, for $i = 1, \ldots, |N|$. The node with the largest $|S_i|$, say node n_l, is first selected as a helper. Then, for each node n_i, for $i = 1, \ldots, |N|$ and $i \neq l$, S_i is updated to remove nodes appearing in S_l, i.e., $\{n_j | n_j \in S_i \wedge n_j \notin S_l\}$ and re-ranked to select the second helper. This procedure repeats until that all nodes are covered or there is no node left.

For the communication overhead of *CoBe*, the main overhead is the required link status information in each beacon including the vehicle ID and the according LoS/NLoS flag of one-hop neighbors. It is totally acceptable, and the analysis process can refer to Sect. 3.4.3.

6.3   Performance Analysis

In this section, we build a two-state Markov model to analyze the performance of *CoBe*, and compare it with the following two benchmark schemes:

- **Conventional 802.11p:** In broadcast mode of 802.11p protocol, vehicles broadcast beacons every 100 ms and there is no rebroadcast mechanisms;
- **Random Forwarding:** In this scheme, a vehicle not only broadcasts beacons as described above but also randomly chooses one or more of its neighbors to rebroadcast its beacons each time.

We analyze the performance of each scheme considering the following two metrics:

- **Beacon Reception Ratio (BRR):** refers to the ratio of beacon reception calculating by the number of neighbors having received the beacon to the total number of one-hop neighbors, which is defined to evaluate the beaconing reliability;
- **Broadcast Utility (BU):** refers to the ratio of the BRR to the total number of broadcast beacons, which is defined to evaluate the rebroadcast cost. For example, if the BRR of 0.9 is achieved when no helper rebroadcasts (i.e., the beacon is broadcasted only once by the transmitter), the BU becomes 0.45 if one more helper is sought (i.e., the beacon is broadcasted twice).

6.3.1   Two-State Markov Chain Model

As the channel switches between LoS and NLoS states, a two-state Markov model can be devised as shown in Fig. 6.1, where the transition probabilities are P_L and P_N, respectively. The likelihood of successfully receiving a packet heavily relies on the current link state, which is P_{good} ($0 < P_{good} < 1$) when the link is in LoS state, and is P_{bad} ($0 < P_{bad} < P_{good}$) when the link is in NLoS state. To further derive variables in this model, we start by stating a known property of the two-state Markov chain:

Fig. 6.1 Two-state Markov
chain model

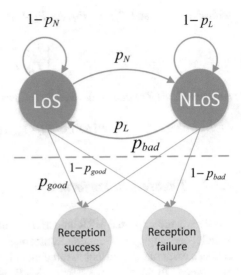

Proposition 6.1 (See, e.g., [15]) *If* $0 < P_L, P_N < 1$, *the unique stationary distribution (or initial state distribution) of the two-state Markov chain in Fig. 6.1 is*

$$\pi = (P_L = \frac{P_L}{P_L + P_N}, P_N = \frac{P_N}{P_L + P_N}),$$

where P_L and P_N (equaling $1 - P_L$) represent the stationary probabilities of the link being in state LoS and NLoS, respectively. Then, we derive the probability $P(Rx)$, that the packet can be successfully received at the kth time slot, $k \geq 1$. According to the law of total probability, $P(Rx)$ can be represented as

$$
\begin{aligned}
P(Rx) &= P(L) \cdot P(Rx|L) + P(N) \cdot P(Rx|N) \\
&= P(L) \cdot P_{good} + P(N) \cdot P_{bad},
\end{aligned}
\tag{6.1}
$$

where $P(L)$ and $P(N)$ is the probability of being in LoS state and in NLoS state, respectively. To obtain the value of $P(L)$ and $P(N)$ at the kth time slot, we can consider all possible unfolding of the Markov chain during k steps. Particularly, at the ith step, $i \leq k$, we denote P_{iL} and P_{iN} as the probability of being LoS state and NLoS state, respectively. According to the unfolding rules of our Markov chain, the recursive relations can be achieved as follows

$$
\begin{cases}
P_{iL} = P_{(i-1)L} \cdot (1 - P_N) + P_{(i-1)N} \cdot P_L \\
P_{iN} = P_{(i-1)L} \cdot P_N + P_{(i-1)N} \cdot (1 - P_L),
\end{cases}
\tag{6.2}
$$

with $P_{1L} = P_L$ and $P_{1N} = P_N$. As $P_L + P_N = 1$, from Eq. (6.2), we can obtain

$$\begin{cases} P_{iL} = P_L \\ P_{iN} = P_N. \end{cases} \tag{6.3}$$

As a result, $P(Rx)$ can be derived as follows

$$P(Rx) = P_L \cdot P_{good} + P_N \cdot P_{bad}. \tag{6.4}$$

6.3.2   Theoretical Performance

We derive the expected BRR and BU of all schemes with respect to one sender vehicle, one single receiver vehicle, and one or multiple helper vehicles. To make the analysis tractable, we further assume that all links are *independently and identically distributed*. In specific, we denote BRR as the probability that one packet can be well received by the receiver and BU as the ratio of BRR to the number of broadcasts.

Performance of Conventional 802.11p As in 802.11p, there is no helper during broadcasting, BRR and BU can be easily obtained as

$$\begin{cases} BRR = P(Rx) \\ BU \;\;= P(Rx). \end{cases} \tag{6.5}$$

In the remainder of this subsection, we derive the results of *CoBe* and random forwarding scheme with one or multiple helpers.

6.3.2.1   Using One Helper

Performance of *CoBe* To derive the BRR of *CoBe*, we divide the analysis into two parts with respective to LoS and NLoS conditions. We can get

$$\begin{aligned} BRR &= P_L \cdot P(BRR|L) + (1 - P_L) \cdot P(BRR|N) \\ &= P_L \cdot P_{good} + (1 - P_L) \cdot (1 - (1 - P_{bad}) \cdot P_{fail}), \end{aligned} \tag{6.6}$$

where P_{fail} means the probability that the retransmission from the helper is unsuccessful. The failure of retransmission can happen with two reasons. One is that the helper does not hear the notification of retransmission from the sender, and the other one is that the retransmitted packet is dropped. Considering n neighbors around the sender and receiver, *CoBe* may choose the helper with the probabilities $p_1, p_2,$ and p_3 in terms of link conditions of sender-to-helper and helper-to-receiver, i.e., p_1 representing the probability that both links are with LoS conditions, p_2

being the probability that one link is with LoS while the other one is with NLoS condition, and p_3 being the probability of both links with NLoS conditions. They can be calculated as follows

$$\begin{cases} p_1 = 1 - (1 - P_L^2)^n \\ p_2 = 1 - p_1 - p_3 \\ p_3 = (1 - P_L)^{2n}. \end{cases} \tag{6.7}$$

Then, P_{fail} can be written as

$$P_{fail} = p_1 \cdot (1 - P_{good}^2) + p_2 \cdot (1 - P_{good} \cdot P_{bad}) + p_3 \cdot (1 - P_{bad}^2). \tag{6.8}$$

To investigate the cost of *CoBe*, we denote the average transmitted packets by the sender and the helper as *num*. Apparently, BU can be written as

$$BU = \frac{BRR \cdot 1}{num} = \frac{BRR}{num}. \tag{6.9}$$

Similarly, we compute *num* in LoS and NLoS two situations, respectively. We can obtain

$$num = P_L \cdot 1 + (1 - P_L) \cdot (1 + PKT_{helper}), \tag{6.10}$$

where PKT_{helper} is the average transmitted packets from the helper. It can be calculated as follows

$$PKT_{helper} = p_1 \cdot P_{good} + \frac{p_2}{2} \cdot (P_{good} + P_{bad}) + p_3 \cdot P_{bad}. \tag{6.11}$$

Then, BU can be derived.

Performance of Random Forwarding For fair comparison in terms of using the same number of helpers with *CoBe*, in this scheme, the sender will always randomly select one helper from its neighbors to retransmit the beacon.

To derive the BRR in this scheme, we consider two events, i.e., the helper does not receive the retransmission notification and the helper receives the notification then retransmits the packet. Then BRR can be calculated as

$$BRR = (1 - P(Rx)) \cdot P(Rx) + P(Rx) \cdot (1 - (1 - P(Rx)^2), \tag{6.12}$$

Similarly, we calculate the average transmitted packets to derive BU. Then BU can be easily obtained as

$$BU = \frac{BRR \cdot 1}{1 + P(Rx) \cdot 1} = \frac{BRR}{1 + P(Rx)}. \tag{6.13}$$

6.3.2.2 Using *H* Helpers

In fact, even with a helper, reception failures may also be encountered due to the severe channel fading. In this case, we conduct analysis of seeking multiple helpers to expand the overlap of covering sets, which can conservatively enhance the reliability.

Performance of *CoBe* When using H helpers, $1 \leq H \leq N$, Eq. (6.6) can also be satisfied. P_{fail} means the probability that the retransmissions from H helpers are unsuccessful simultaneously. It can be written as

$$P_{fail} = \Pi_{i=1}^{H} p_i, \tag{6.14}$$

where p_i for $1 \leq i \leq H$, is the probability of unsuccessful retransmission from the ith helper. The value of p_i depends on the link conditions from the sender to the ith helper and from the ith helper to the receiver. The two links can be in LoS-LoS, LoS-NLoS/NLoS-LoS, or NLoS-NLoS three categories. We define P_{LL}, P_{NL}, and P_{NN} as the probability of unsuccessful retransmission from the helper under link conditions of three respective categories. Thus, they can be achieved as follows

$$\begin{cases} P_{LL} = 1 - P_{good}^2 \\ P_{NL} = 1 - P_{good} \cdot P_{bad} \\ P_{NN} = 1 - P_{bad}^2. \end{cases} \tag{6.15}$$

In addition, given two links, they can be LoS-LoS, LoS-NLoS/NLoS-LoS, and NLoS-NLoS conditions with the probability of P^{LL}, P^{NL}, and P^{NN}, respectively. They can be obtained as

$$\begin{cases} P^{LL} = P_L^2 \\ P^{NL} = 1 - P_L^2 - (1 - P_L)^2 \\ P^{NN} = (1 - P_L)^2. \end{cases} \tag{6.16}$$

For H helpers, we define the event that i helpers, j helpers, and $H - i - j$ helpers are with link conditions of LoS-LoS, LoS-NLoS/NLoS-LoS, and NLoS-NLoS, respectively; the event happens with the probability of $P(i, j)$ for $i, j, H-i-j \geq 0$. Then, Eq. (6.14) can be replaced as

$$P_{fail} = \Sigma P(i, j) \cdot P_{LL}^i \cdot P_{NL}^j \cdot P_{NN}^{(H-i-j)}, i, j \geq 0, i + j \leq H. \tag{6.17}$$

The value $P(i, j)$ depends on the rule of helper choosing. According to *CoBe* design, $P(i, j)$ can be achieved as follows

$$P(i, j) = \begin{cases} \Sigma_{m=H}^{n} C_n^m (P^{LL})^m (1 - P^{LL})^{n-m}, \\ \qquad\qquad i = H; \\ \Sigma_{m=H-i}^{n-i} C_n^i C_{n-i}^m (P^{LL})^i (P^{NL})^m (P^{NN})^{n-i-m}, \\ \qquad\qquad 0 \le i < H, j = H - i; \\ C_n^i C_{n-i}^j (P^{LL})^i (P^{NL})^j (P^{NN})^{n-i-j}, \\ \qquad\qquad 0 \le i < H, 0 \le j < H - i. \end{cases} \tag{6.18}$$

Then BRR can be derived. To calculate BU in this condition, Eqs. (6.9) and (6.10) can also be utilized. However, the derivation of PKT_{helper} is different from Eq. (6.11), which follows

$$PKT_{helper} = \Sigma P(i, j) \cdot (i \cdot P_{good} + \frac{j}{2} \cdot (P_{good} + P_{bad}) \tag{6.19}$$
$$+ (N - i - j) \cdot P_{bad}).$$

Then, BU can be derived.

Performance of Random Forwarding For H helpers, we define the event that i helpers for $i \in [0, H]$ receive the retransmission notification and retransmits the packet, while the remaining $H - i$ helpers do not receive the retransmission notification. The event happens with a probability $P(i)$. Then, BRR can be calculated as

$$BRR = \Sigma_{i=0}^{H} P(i) \cdot (1 - (1 - P(Rx))^{i+1}), 0 \le i \le H, \tag{6.20}$$

and BU can be obtained by

$$BU = \frac{BRR}{1 + \Sigma_{i=0}^{H} P(i) \cdot i}, 0 \le i \le H. \tag{6.21}$$

In addition, the probability $P(i)$ follows

$$P(i) = C_H^i \cdot P(Rx)^i \cdot (1 - P(Rx))^{H-i}, 0 \le i \le H. \tag{6.22}$$

6.4 Synthesizing V2V Communication Traces

Although we have collected a large volume of V2V traces for a pair of vehicles, they are still not enough for extensive trace-driven experiments when involving more vehicles in the network. In addition, as collecting IEEE 802.11p V2V communication trace is labor-intensive and time-consuming, in this chapter, we propose a scheme to synthesize abundant V2V traces based on our previously found distribution results, which can also be of significant importance for practical analysis and model validation. Specifically, we adopt the following three procedures to do it:

1. As both LoS and NLoS durations follow a power law distribution, we utilize the power law equation to fit the empirical data and then output sets of LoS and NLoS duration values.
2. Within LoS duration, we adopt the exponential distribution equation to fit empirical PIR times, which can generate a set of PIR times within the duration. For a PIR time, the packets at the start and the end slots are well received while other packets during the time are lost.
3. Likewise, within NLoS duration, we utilize the exponential distribution equation to fit empirical PIL times with generating a set of PIL times within the duration. In contrast, given a PIL time, the packets at the start and the end points are lost while other packets during the time are well received.

In the following subsections, we first detail the three procedures to fit and generate related data, and then combine them together to output the desired communication trace data.

6.4.1   Fitting LoS and NLoS Durations

According to our measurement study in the previous chapter that LoS durations follow a power law distribution, we can utilize the general mode

$$P(LoS \quad duration > x) = C \cdot x^{-a}, \tag{6.23}$$

to fit the CCDF of LoS durations, where x satisfies $x \geq 2$ in accordance with the collected empirical data. Fitting metrics of SSE, R-square, Adjusted R-square, and RMSE are evaluated, and Table 6.1 shows the fitting results, where precise fitting

Table 6.1   Fitting results of LoS durations in different environments

Fitting results	Suburban	Highway	Urban
Coefficient C (with 95% confidence bounds)	1.098 (1.092, 1.103)	1.014 (1.01, 1.019)	1.081 (1.073, 1.089)
Coefficient a (with 95% confidence bounds)	0.3768 (0.3751, 0.3784)	0.3771 (0.3758, 0.3785)	0.3293 (0.327, 0.3316)
SSE	0.0008859	0.0004907	0.001941
R-square	0.9993	0.9996	0.9985
Adjusted R-square	0.9993	0.9996	0.9985
RMSE	0.003086	0.002273	0.00467

results cross verify our findings in the data analytics. The CDF of LoS durations then satisfies

$$P(LoS \quad duration \leq x) = 1 - C \cdot x^{-a}. \tag{6.24}$$

In our case, x can take only a discrete set of integer values, e.g., $x = 2, 3, 4, \ldots, N$. According to the fitting results, for each environment, a probability set can be computed, denoted as $\mathcal{P} = [p_2, p_3, \ldots, p_N]$ where p_i for $i \in [2, N]$ is the value of $P(LoS \quad duration \leq i)$. Therefore, the probability of duration $x = i$ is $p_i - p_{i-1}$. Specifically, when $i = 2$, the value of p_{i-1} is 0. To synthesize an LoS duration, we randomly produce a variable in the interval $[0, 1]$; if the variable locates between $(p_{i-1}, p_i]$, then the duration is the value i.

Following the power law distribution of the NLoS durations, we utilize the general model

$$P(NLoS \quad duration > x) = C \cdot x^{-a}, \tag{6.25}$$

to fit the CCDF of NLoS durations in different environments, and the results are demonstrated in Table 6.2. Then the CDF of NLoS durations satisfies

$$P(NLoS \quad duration \leq x) = 1 - C \cdot x^{-a}. \tag{6.26}$$

Differently, according to the collected empirical data, the NLoS duration x can take a discrete set of integer values starting from 1, i.e., $x = 1, 2, 3, \ldots, N$. This is also the main reason that NLoS durations are in general shorter than LoS durations, as the probability of NLoS duration being 1 s is about 0.351, 0.354 and 0.435 in suburban, highway, and urban, respectively. Based on the fitting results, a set denoted as $\mathcal{P} =$

Table 6.2 Fitting results of NLoS durations in different environments

Fitting results	Suburban	Highway	Urban
Coefficient C (with 95% confidence bounds)	0.6556 (0.6522, 0.659)	0.663 (0.6483, 0.6776)	0.5807 (0.5714, 0.59)
Coefficient a (with 95% confidence bounds)	0.4504 (0.4477, 0.4531)	0.6574 (0.6427, 0.6721)	0.4757 (0.4684, 0.4831)
SSE	0.0001191	0.0019	0.001702
R-square	0.9997	0.996	0.9959
Adjusted R-square	0.9997	0.9958	0.9959
RMSE	0.002063	0.008238	0.005954

$[p_1, p_2, \ldots, p_N]$ where p_i for $i \in [1, N]$ is the value of $P(NLoS \quad duration \leq i)$, can be calculated. Likewise, the NLoS durations can be synthesized.

6.4.2   Fitting PIR Times Under LoS Conditions

As the PIR times follow an exponential distribution under LoS conditions, we adopt the general model

$$P(PIR > x) = e^{-\lambda \cdot x}, \tag{6.27}$$

to fit the CCDF of PIR times under LoS conditions, and the results are shown in Table 6.3. According to the empirical data, the PIR time x can take a discrete set of values, i.e., $x = 0.1, 0.2, 0.3, \ldots$[1] Then, the CDF of PIR times satisfies

$$P(PIR \leq x) = 1 - e^{-\lambda \cdot x}. \tag{6.28}$$

Based on the coefficient λ in Table 6.3, we achieve a set $\mathcal{P} = [p_1, p_2, \ldots, p_N]$ where p_i for $i \in [\frac{1}{10}, \frac{N}{10}]$ is the value of $P(PIR \leq \frac{i}{10})$. To synthesize a PIR time under LoS conditions, we first randomly produce a variable within $[0, 1]$; if the variable locates within $(p_{i-1}, p_i]$, then the time is the value $\frac{i}{10}$. Given an LoS duration T, we can synthesize a PIR time list, denoted as $\mathcal{PIR} = [PIR_1, PIR_2, \ldots, PIR_K]$ where PIR_i for $i \in [1, K]$ is the ith generated PIR time, and the total time $\sum_{i=1}^{K} PIR_i$ is the value T.

6.4.3   Fitting PIL Times Under NLoS Conditions

Likewise, we adopt the same process to synthesize PIL times under NLoS conditions. A generate model

Table 6.3 Fitting results of PIR times under LoS conditions in different environments

Fitting results	Suburban	Highway	Urban
Coefficient λ (with 95% confidence bounds)	40.31 (39.78, 40.85)	34.63 (34.15, 35.11)	38.41 (37.99, 38.83)
SSE	1.357e−06	4.537e−06	1.677e−06
R-square	0.995	0.9947	0.9959
Adjusted R-square	0.995	0.9947	0.9959
RMSE	0.0004119	0.0006736	0.0004095

---

[1]Note that, as the beacon intervals are set to 0.1 s in the data collection campaigns (the negligible transmission time is omitted), the PIR time can be the multiples of 0.1 s in empirical data set.

Table 6.4 Fitting results of PIL times under NLoS conditions in different environments

Fitting results	Suburban	Highway	Urban
Coefficient λ (with 95% confidence bounds)	18.04 (17.24, 18.84)	21.78 (21.19, 22.37)	26.09 (25.38, 26.8)
SSE	19.51e−05	7.187e−05	4.215e−05
R-square	0.9905	0.9933	0.9906
Adjusted R-square	0.9905	0.9933	0.9906
RMSE	0.005702	0.002997	0.002295

$$P(PIL > x) = e^{-\lambda \cdot x}, \tag{6.29}$$

is utilized to fit CCDF of PIL times under NLoS conditions, and the results are shown in Table 6.4. The CDF of PIL time then satisfies

$$P(PIL \leq x) = 1 - e^{-\lambda \cdot x}. \tag{6.30}$$

Computing an set $\mathcal{P} = [p_1, p_2, \ldots, p_N]$ where p_i for $i \in [\frac{1}{10}, \frac{N}{10}]$ is the value of $P(PIL \leq \frac{i}{10})$, then a PIL time list $\mathcal{PIL} = [PIL_1, PIL_2, \ldots, PIL_K]$ can be generated, where the total time $\sum_{i=1}^{K} PIL_i$ is the given NLoS duration T.

6.4.4 Outputting Link Communication Traces

For a communication Link, the LoS and NLoS conditions appear alternately. As shown in Fig. 6.2c, we first generate an LoS duration list and an NLoS duration list, denoted as $\mathcal{LS} = [LoS_1, LoS_2, \ldots, LoS_n]$ and $\mathcal{NLS} = [NLoS_1, NLoS_2, \ldots, NLoS_n]$, respectively. For each LoS_i for $i \in [1, n]$, we generate a PIR time list $\mathcal{PIR}$. For each PIR time, the packets at the start and the end slots are well received while other packets during the time are lost. As shown in Fig. 6.2a, it is a sequence of packets sent in the duration of LoS_i, where white blocks denote successfully received packets while dark blocks denote packet reception failures. PIR_1, PIR_2, and PIR_3 is the example of PIR time being $1X$, $2X$, and $3X$ of beacon intervals, respectively. Likewise, for each $NLoS_i$ for $i \in [1, n]$, we generate a PIL time list $\mathcal{PIL}$. In contrast, for each PIL time, the packets at the start and the end slots are lost but other packets during the time are well received. As shown in Fig. 6.2b, it is a sequence of packets sent in the duration of $NLoS_i$. PIL_1, PIL_2, and PIL_3 is the example of PIL time being $1X$, $2X$, and $3X$ of beacon intervals, respectively. In doing so, we can synthesize abundant V2V trace by customizing the value of n, i.e., the length of LoS and NLoS duration list.

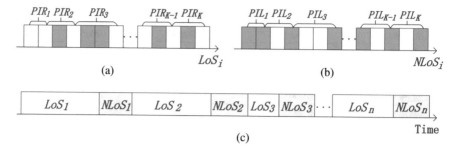

Fig. 6.2 The data presentation of V2V communication trace. (**a**) A PIR time list under a LoS duration. (**b**). A PIL time list under a NLoS duration (**c**) Alternate LoS and NLoS durations

6.5   Performance Evaluation

In this section, we carry out the performance evaluation, where the NLoS detection accuracy is first investigated, and then the beaconing performance of *CoBe* is evaluated in terms of BRR and BU.

6.5.1   NLoS Detection Accuracy

To evaluate NLoS detection accuracy, we adopt the cross-validation scheme to examine the performance of different machine learning algorithms. To be specific, for the **trace** $\mathcal{H}$, $\mathcal{S}$ and $\mathcal{U}$ with respective 16,425, 16,033 and 27,439 samples of labeled LoS/NLoS conditions, we first divide them into 10 equal-size subsets. Then, for each round experiment, we choose 1 subset as the testing set and aggregate the other 9 subsets as the training set. Based on the testing cases of True Positive (TP), True Negative (TN), False Positive (FP), and False Negative (FN), the following five metrics are evaluated:

- *Accuracy:* refers to the probability that a condition is correctly identified, i.e., $\frac{TP+TN}{TP+FP+TN+FN}$;
- *Precision:* refers to the probability that an identified NLoS conditions is correctly identified, i.e., $\frac{TP}{TP+FP}$;
- *Recall:* refers to the probability that all NLoS conditions in ground truth are correctly identified, i.e., $\frac{TP}{TP+FN}$;
- *F-Score:* is calculated by combining the precision and recall metric together, i.e., $2 \times \frac{Precision \times Recall}{Precision+Recall}$;
- *False Positive Rate (FPR):* refers to the probability that an LoS condition is identified as an NLoS condition, i.e., $\frac{FP}{FP+TN}$.

Precise NLoS Detection   We plot the average scores of F-Score and FPR under different environments in Fig. 6.3, where we can observe that NLoS conditions

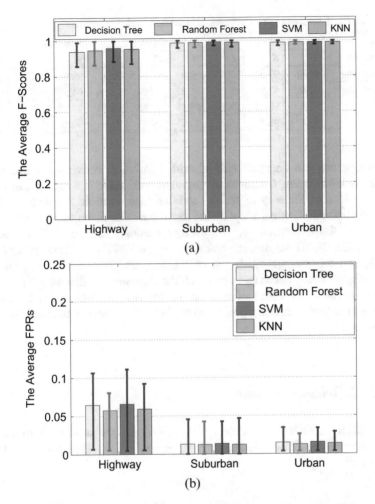

Fig. 6.3 NLoS detection performance. (**a**) The average F-Scores. (**b**) The average FPRs

can be well identified by machine learning algorithms, especially for SVM and KNN algorithms. For instance, as shown in Fig. 6.3a, the SVM algorithm can achieve the average F-Scores of 95.9%, 99.1%, and 99.2% in respective highway, urban, and suburban environments, and the value is about 95.5%, 99%, and 99.1% when adopting the KNN algorithm. The detection results demonstrate that NLoS conditions can be precisely identified in real time. On the contrary, the scores of average FPRs can be guaranteed to be rather small by adopting the two algorithms. For example, when adopting the SVM algorithm, the average FPR is only 6.6%, 1.5%, and 1.4% in respective highway, urban, and suburban environments. It should be pointed out that higher FPRs are usually triggered in highway environments, since packets can be accidentally dropped in highways due to the fast speed of

Table 6.5 NLoS detection results achieved by machine learning algorithms

Algorithms	Accuracy (%)			Precision (%)			Recall (%)		
	$\mathcal{H}$	$\mathcal{S}$	$\mathcal{U}$	$\mathcal{H}$	$\mathcal{S}$	$\mathcal{U}$	$\mathcal{H}$	$\mathcal{S}$	$\mathcal{U}$
Decision tree	90.86	97.52	97.57	94.39	98.71	98.77	93.68	98.74	98.58
Random forest	92.08	97.91	98.08	95.22	99.04	99.16	94.37	98.83	98.77
SVM	93.67	98.31	98.28	98.11	99.65	99.63	93.93	98.67	98.51
KNN	93.0	98.15	98.21	96.6	99.31	99.4	94.34	98.82	98.67

vehicles, which can be mistakenly identified as NLoS conditions. Other metric results under different environments are shown in Table 6.5, which can further demonstrate the efficiency of NLoS condition identification. For instance, under urban environments, when adopting the algorithm of decision tree, random forest, SVM, and KNN, respectively, the average accuracy can be 97.57%, 98.08%, 98.28%, and 98.21, the average precision can be 98.77%, 99.16%, 99.63%, and 99.4%, and the average recall can be 98.58%, 98.77%, 98.51%, and 98.67%. Moreover, we can observe that among all the machine learning algorithms, SVM is able to achieve the best performance. To this end, in *CoBe*, we will adopt the SVM as the NLoS detection algorithm in the following experiments for performance comparison.

6.5.2   Efficiency of CoBe

We conduct extensive trace-driven experiments in this subsection to evaluate the performance of *CoBe*, where both the numerical and experiment results are presented.

6.5.2.1   Using One Helper

We first examine the impact of the number of neighbors on the relaying decision, where only one helper is used.

Experiment Setup   Based on the fitting results of **trace** $\mathcal{U}$, 100 link traces are first synthesized, and each trace lasts 1000 min. For each round of experiment, we first randomly choose two link traces to represent the sending/receiving process between the sender and receiver. To include a neighbor vehicle, we choose another two pairs of link traces to represent the sending/receiving process between the sender and neighbor, and between the neighbor and receiver. The number of neighbors n is ranged from 1 to 10, and we conduct the experiment over each value. We set other parameters according to the fitting results of **trace** $\mathcal{U}$, where P_L and P_N are set to 0.8 and 0.2, respectively, and P_{good} and P_{bad} are respectively set to 0.97 and 0.3 in accordance with the average PDR under LoS and NLoS conditions. For each setting, 30 rounds of experiments are conducted in order to achieve the average results with statistically significant. Detailed experiment parameters are shown in Table 6.6.

Performance Comparison We plot the average BRRs achieved by different beaconing strategies in Fig. 6.4a, where the numerical and experiment results are respectively presented by dashed and solid lines. We can observe that the experiment and numerical results have similar variation trends, where we can make two major statements. First, the beaconing reliability can be significantly enhanced with seeking one helper to rebroadcast the beacons. For instance, in 802.11p, the numerical and experiment BRR is about 83.6% and 85%, but when adopting *CoBe* and the random approach, the BRR can be respectively enhanced to 96% and 95%

Table 6.6 Experiment parameters

Parameters	Value	Parameters	Value
Channel number	178	Neighbors n	[1, 10] or 10
Channel bandwidth	10 MHz	Number of links	100
Transmission power	14 dBm	Chosen links	[6, 42] or 42
Data rate	3 Mbps	P_L	0.8
Environment	Urban	P_N	0.2
Number of lanes	4–8	P_{good}	0.97
Mean distance	147.796 m	P_{bad}	0.3
Mean Tx speed	5.107 m/s	Simulation time	1000 min
Mean Rx speed	5.041 m/s	Simulation rounds	30
Helpers H	1 or [1, 10]		

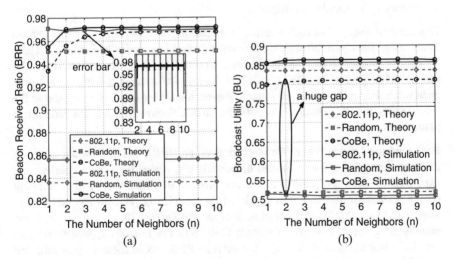

Fig. 6.4 The performance of using one helper. (**a**) BRR vs. the number of neighbors. (**b**) BU vs. the number of neighbors

for numerical results, and to 97.1% and 96.8% for experiment results. Second, a better performance can be achieved by *CoBe* when there are more helper selections (i.e., increasing the number of neighbors), but the random approach and 802.11p cannot react to the environment. For instance, when increasing the number of neighbors, the experiment results of BRRs keep constant and fluctuate within a small range in 802.11p and the random approach, respectively. However, the BRRs can increase gradually when adopting *CoBe*. In addition, we can observe that when with more than two neighbors, *CoBe* can achieve the BRR up to 97%, which is higher than 96.8% that achieved by the random approach. Moreover, the error bar of two schemes can further demonstrate the stability of *CoBe*.

We plot the average BUs in Fig. 6.4b, where we can observe that the lowest broadcast utility (below 52%) for both the numerical and experiment results, generally appear in the random approach. The value is rather low when compared with other two strategies with the utility above 80%, which can inflict significant overhead to the system. Besides, we can observe that when with more neighbors, *CoBe* can achieve a better performance of BU. For instance, in terms of experiment results, the BU can gradually increase and reach above 86.2% when with more than two neighbors, which can outperform the 802.11p with the BU of 85.6%.

To summarize, compared with two benchmarks, *CoBe* can enhance the BRR significantly with a negligible BU degradation, but when with more neighbors, *CoBe* can achieve the supreme performance in all aspects.

6.5.2.2   Using *H* Helpers

We then examine the upper-bound of beaconing reliability that *CoBe* can achieve, by increasing the number of helper *H*.

Experiment Setup   For each round of experiment, two link traces are first chosen to represent the sending/receiving process between the sender and receiver. Then, we choose another 20 pairs of link traces to inject 10 neighbors into the system. The number of helpers *H* is ranged from 1 to 10, and we conduct the experiment over each value. For each setting, 30 rounds of experiments are conducted to achieve the average results with statistically significant. Table 6.6 shows the detailed experiment parameters.

Performance Comparison   We plot the average BRRs and BUs when adopting different strategies in Fig. 6.5a and b, respectively, where both the numerical and experiment results are presented. We can observe that numerical and experiment results present similar variation trends. More specifically, without any retransmission, 802.11p achieves the constant average BRRs and BUs when varying the number of helpers. However, for both *CoBe* and the random approach, we can see that when using more helpers, the average BRR would increase gradually but the BU decreases significantly. For instance, when adopting *CoBe*, the average numerical BRR increases from 96.8% to 97.6% but the average numerical BU

Fig. 6.5 The performance of using H helpers. (**a**) BRR vs. the number of helpers. (**b**) BU vs. the number of helpers

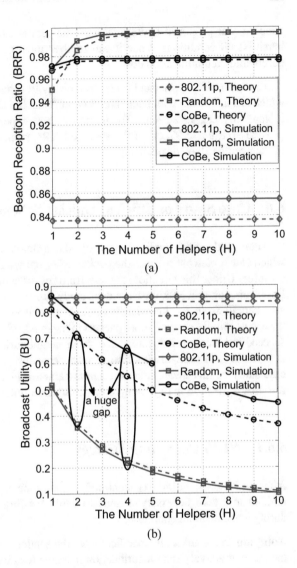

(a)

(b)

can decrease from 81.1% to 36.5% when using more helpers. Likewise, when adopting the random approach, the average numerical BRR increases from 95.1% to 100% with the average numerical BU decreasing from 51.8% to 10.7%. With the performance results, we can conclude that to enhance the communication reliability to an extreme level, e.g., above 97%, it means a huge cost to be paid, especially for the random approach. Although the random approach can push forward the upper-bound reliability to 100% when seeking more than 4 helpers, a dramatic performance degradation of BUs would appear, which can result in a performance gap over 30% when compared with that of *CoBe*. It can significantly aggravate the resource shortage problem in VANETs [16–19]. This inspires us that when adopting our proposed *CoBe*, seeking 2 helpers is usually enough when meeting harsh NLoS conditions, since it can recover most of the packet failures with an acceptable communication cost.

To summarize, *CoBe* can be a smart strategy to enhance the beaconing reliability which can ably cope with harsh NLoS conditions. In most cases, seeking one helper is generally suggested since for numerical and experiment results, the case can provide 96.8% and 97.2% of the average BRRs, respectively, and 81% and 86.2% of the average BUs, respectively.

6.6   Case Study of Efficient Unicast Scheme

For non-road-safety applications [20], the unicast scheme is usually adopted, which can somewhat tolerate the packet delay but would cost substantial spectrum resource. Under this situation, the spectrum resource utilization is more important. Based on our previous measurement study, upper-layer *unicast* applications can also benefit from the knowledge of current environmental context. For instance, instead of blindly sending more packets in harsh NLoS conditions, which can hardly succeed but increase interferences to other neighbouring vehicles, a better strategy is to avoid data transmission during such occasions. In this section, we present a case study of link-aware unicast scheme named *Bungee*, to improve the spectrum resource utilization in regarding to the LoS/NLoS link conditions.

6.6.1   NLoS Unicasting Strategy

After an NLoS condition is identified at the receiver side, the sender should be notified and take necessary actions to avoid wasting efforts in sending packets during a bad channel condition.

Notifying the Sender In specific, to get the sender notified, one solution is to let the receiver actively send a notification message (or piggyback such information in its own beacon packets towards the sender). Since notification packets are sent after NLoS conditions are detected, the probability of such notification packets being dropped is high. A candidate solution is to let the sender detect the channel quality between the receiver and itself (e.g., using physical layer hints or collecting packets sent from the receiver to calculate PDR), leveraging the high symmetric correlation between the incoming and outgoing links in vehicular networks [3].

Avoiding Data Transmission in NLoS Conditions For non-road-safety applications, given an identified NLoS condition caused by dynamic vehicle traffic, for the transmitter, one simple strategy is to wait until the channel condition gets better.

As the tail distributions of PIL times have exponential decays in NLoS conditions across all environments, the number of slots between two consecutive packet reception failures also follows an exponential distribution, which implies the probability of short PIL times is relatively high. For example, as shown in Fig. 5.14, the probability of observing two consecutive packet reception failures in suburban, highway and urban environment is 87.8%, 92.6% and 96.7%, respectively. In addition, such probability in different NLoS conditions in the same environment also varies. For example in Fig. 5.17, the probability of observing two consecutive packet reception failures in severe, intermediate and normal NLoS condition is about 95%, 70% and 40%, respectively.

Based on these observations, one superior scheme to avoid transmitting data in NLoS conditions should be able to adaptively jump the duration of a particular NLoS condition. We present one such scheme, called *Bungee*, based on our statistical analysis results. Specifically, given the exponential distribution of PIL times, the random variable of PIL time is memoryless. Let $1 - p$ denote the probability of observing one packet reception failure and the probability of observing k consecutive packet reception failures P_k is

$$P_k = (1 - p)^k. \tag{6.31}$$

After the sender notices an NLoS condition in its outgoing link, it omits to transmit beacons during the next T message slots if the probability of T consecutive packet reception failures is higher than a jumping threshold α,

$$P_T = (1 - p)^T \geq \alpha. \tag{6.32}$$

As illustrated in Fig. 6.6, the sender calculates the PDR with a larger circular checking queue of size Q as a better approximation of p and jumps over the maximum $T = \frac{\ln \alpha}{\ln(1-p)}$ message slots. After that, it empties the circular checking queue and repeats the whole procedure. Note that the appropriate α in different urban environments can be trained with history data. Note that it is possible that the estimated NLoS duration is longer than the ground truth, which leads to valid transmission slots in the beginning of the next LoS condition being wasted. For example in Fig. 6.6, W slots are wasted. We also evaluate the performance of this scheme in the next subsection.

6.6.2 Performance Evaluation

In this subsection, we conduct extensive trace-driven simulations across all environments to evaluate the performance of *Bungee*. We divide all traces into two parts, one for *training* and the other for *testing*.

We examine the *performance gain*, defined as the ratio of successfully jumped NLoS durations to the whole duration of an NLoS condition. In addition, we examine the *performance waste*, defined as the ratio of the number of wasted LoS

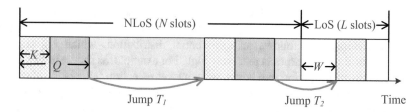

Fig. 6.6 The sender uses a queue of K slots to fast detect an NLoS condition; if an NLoS is identified, it uses another queue of Q slots to estimate PDR, and avoid to transmit packets in the next calculated T slots

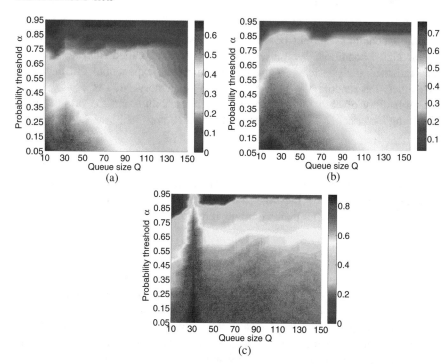

Fig. 6.7 Performance gain in all environments. (**a**) Suburban. (**b**) Highway. (**c**) Urban

slots to the whole duration of the corresponding LoS condition. For example in Fig. 6.6, the performance gain is calculated as $(T_1 + T_2 - W)/N$ and the waste is calculated as W/L. We vary the online learning queue size Q from 10 slots to 100 slots with an interval of 5 slots, vary the jumping threshold α from 0.05 to 0.95 with an interval of 0.05, and conduct a cross verification for the optimal performance gain and waste across all training traces.

Figures 6.7 and 6.8 shows the average performance gain and waste as a function of the queue size Q and probability threshold α over all testing traces. It can be seen that with the optimal configuration of $Q = 20$ and $\alpha = 0.05$, *Bungee* can achieve

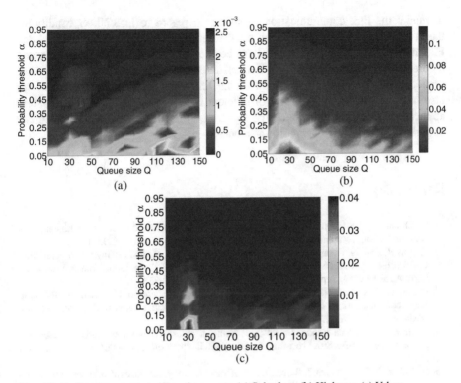

Fig. 6.8 Performance waste in all environments. (**a**) Suburban. (**b**) Highway. (**c**) Urban

the best performance gain of 67.6% and 76.1% in suburban and highway scenarios respectively. For the urban scenario, the optimal performance gain of 87.7% can be achieved with $Q = 30$ and $\alpha = 0.05$. Although the corresponding performance waste is not the lowest, it is a small value of 2.4%, 9.2%, and 4% in suburban, highway and urban, respectively, which is acceptable. Another observation is that, with wide ranges of Q and α, e.g., $10 \leq Q \leq 50$ and $0.05 \leq \alpha \leq 0.45$, *Bungee* can still obtain a good performance gain about 60%, 70%, and 80% with a relatively low performance waste about 2%, 10%, and 4% in suburban, highway, and urban respectively. We can conclude that *Bungee* can work robustly due to its low precision requirement for parameter configuring of Q and α.

6.7   Summary

In this chapter, we have investigated the link-aware reliable broadcasting scheme design for road-safety applications. Particularly, we have proposed *CoBe* to enhance the broadcast reliability by coping with harsh NLoS conditions, which integrates three major components: (1) online NLoS detection; (2) link status exchange; and (3) beaconing with helpers. Additionally, we have built a two-state Markov chain to model the link communication behaviors in regard to LoS/NLoS conditions, based on which we have carried out the theoretical performance analysis of *CoBe*. Extensive trace-driven simulations have been conducted, and both simulation and theoretical numerical results have demonstrated the efficacy of *CoBe* in terms of broadcasting reliability. In addition to the link-aware broadcasting scheme, we also have presented a case study of efficient unicast scheme for non-road-safety applications, which can enhance the spectrum utilization by avoiding inadequate opportunities for data transmission in advance.

References

1. W. Zhuang, Q. Ye, F. Lyu, N. Cheng, J. Ren, SDN/NFV-empowered future IoV with enhanced communication, computing, and caching. Proc. IEEE **108**(2), 274–291 (2020)
2. F. Lyu, H. Zhu, N. Cheng, H. Zhou, W. Xu, M. Li, X. Shen, Characterizing urban vehicle-to-vehicle communications for reliable safety applications. IEEE Trans. Intell. Transp. Syst. 1–17, Early Access (2019). https://doi.org/10.1109/TITS.2019.2920813
3. F. Bai, D.D. Stancil, H. Krishnan, Toward understanding characteristics of dedicated short range communications (DSRC) from a perspective of vehicular network engineers, in *Proceedings of ACM MobiCom*, Sept 2010
4. F. Martelli, M. Elena Renda, G. Resta, P. Santi, A measurement-based study of beaconing performance in IEEE 802.11p vehicular networks, in *Proceedings of IEEE INFOCOM*, May 2012
5. M. Boban, T.T.V. Vinhoza, J. Barros, M. Ferreira, O.K. Tonguz, Impact of vehicles as obstacles in vehicular ad hoc networks. IEEE J. Sel. Areas Commun. **29**(1), 15–28 (2011)
6. H. Zhou, N. Cheng, N. Lu, L. Gui, D. Zhang, Q. Yu, F. Bai, X. Shen, WhiteFi infostation: engineering vehicular media streaming with geolocation database. IEEE J. Sel. Areas Commun. **34**(8), 2260–2274 (2016)
7. F. Lyu, J. Ren, N. Cheng, P. Yang, M. Li, Y. Zhang, X. Shen, LEAD: large-scale edge cache deployment based on spatio-temporal WiFi traffic statistics. IEEE Trans. Mob. Comput. 1–16 (2020). https://doi.org/10.1109/TMC.2020.2984261
8. Y.X. Xie, Z.L. Li, M. Li, Precise power delay profiling with commodity WiFi, in *Proceedings of ACM MobiCom*, Sept 2015
9. J. Liang, Z. Qin, S. Xiao, L. Ou, X. Lin, Efficient and secure decision tree classification for cloud-assisted online diagnosis services. IEEE Trans. Dependable Secure Comput. 1–13 (2019). https://doi.org/10.1109/TDSC.2019.2922958
10. H. Peng, X. Shen, Deep reinforcement learning based resource management for multi-access edge computing in vehicular networks. IEEE Trans. Netw. Sci. Eng. (2020). https://doi.org/10.1109/TNSE.2020.2978856
11. H.A. Omar, W. Zhuang, L. Li, VeMAC: a TDMA-based MAC protocol for reliable broadcast in VANETs. IEEE Trans. Mob. Comput. **12**(9), 1724–1736 (2013)
12. H. Zhou, W. Xu, J. Chen, W. Wang, Evolutionary V2X technologies toward the internet of vehicles: challenges and opportunities. Proc. IEEE **108**(2), 308–323 (2020)
13. F. Lyu, H. Zhu, H. Zhou, L. Qian, W. Xu, M. Li, X. Shen, MoMAC: mobility-aware and collision-avoidance MAC for safety applications in VANETs. IEEE Trans. Veh. Technol. **67**(11), 10590–10602 (2018)
14. Z. Zhang, J. Willson, Z. Lu, W. Wu, X. Zhu, D. Du, Approximating maximum lifetime k-coverage through minimizing weighted k-cover in homogeneous wireless sensor networks. IEEE/ACM Trans. Netw. **24**(6), 3620–3633 (2016)
15. A.E.F. Clementi, C. Macci, A. Monti, F. Pasquale, R. Silvestri, Flooding time of edge-Markovian evolving graphs. SIAM J. Discrete Math. **24**(4), 1694–1712 (2010)

16. F. Lyu, N. Cheng, H. Zhu, H. Zhou, W. Xu, M. Li, X. Shen, Towards rear-end collision avoidance: adaptive beaconing for connected vehicles. IEEE Trans. Intell. Transp. Syst. 1–16, Early Access (2020). https://doi.org/10.1109/TITS.2020.2966586

17. H. Peng, Q. Ye, X. Shen, Spectrum management for multi-access edge computing in autonomous vehicular networks. IEEE Trans. Intell. Transp. Syst. 1–12 (2019). https://doi.org/10.1109/TITS.2019.2922656

18. X. Cheng, L. Yang, X. Shen, D2D for intelligent transportation systems: a feasibility study. IEEE Trans. Intell. Transp. Syst. 16(4), 1784–1793 (2015)

19. H. Zhou, N. Zhang, Y. Bi, Q. Yu, X. Shen, D. Shan, F. Bai, TV white space enabled connected vehicle networks: challenges and solutions. IEEE Netw. 31(3), 6–13 (2017)

20. N. Cheng, F. Lyu, W. Quan, C. Zhou, H. He, W. Shi, X. Shen, Space/aerial-assisted computing offloading for IoT applications: a learning-based approach. IEEE J. Sel. Areas Commun. 37(5), 1117–1129 (2019)

Chapter 7
Safety-Aware and Distributed Beacon Congestion Control

After enhancing the beaconing performance at both the MAC layer and link layer, we turn to the network layer performance in this chapter. Particularly, under dynamic traffic conditions, especially for dense-vehicle scenarios, the naive beaconing scheme where vehicles broadcast beacons at a fixed rate with a fixed transmission power can cause severe channel congestion and thus degrade the beaconing reliability. In this chapter, by considering the kinematic status and beaconing rate together, we study the rear-end collision risk and define a danger coefficient ρ to capture the danger threat of each vehicle being in the rear-end collision. In specific, we propose a fully distributed *a*daptive *b*eacon *c*ontrol scheme, named *ABC*, which makes each vehicle actively adopt a minimal but sufficient beaconing rate to avoid the rear-end collision in dense scenarios based on individually estimated ρ. With *ABC*, vehicles can broadcast at the maximum beaconing rate when the channel medium resources are enough and meanwhile keep identifying whether the channel is congested. Once a congestion event is detected, an NP-hard distributed beacon rate adaptation (DBRA) problem is solved with a greedy heuristic algorithm, in which a vehicle with a higher ρ will be assigned with a higher beaconing rate while keeping the total required beaconing demand lower than the channel capacity. We prove the efficiency of the heuristic algorithm theoretically, which can achieve the near-optimal result. By using Simulation of Urban MObility (SUMO)-generated vehicular traces, we conduct extensive simulations to demonstrate the efficacy of our proposed *ABC* scheme. Simulation results show that vehicles can adapt beaconing rates in accordance with the road safety demand, and the beaconing reliability can be guaranteed even under high-density vehicle scenarios.

© Springer Nature Switzerland AG 2020
F. Lyu et al., *Vehicular Networking for Road Safety*, Wireless Networks,
https://doi.org/10.1007/978-3-030-51229-3_7

7.1 Problem Statement

With periodically broadcasting beacons, each vehicle can share the location, heading direction, braking status, velocity, etc., with neighboring vehicles in proximity, which can enable many advanced road road-safety applications such as stop sign violation, intersection collision warning, emergency electronic brake lights, and among others [1–3]. However, it is quite challenging to design an efficient and reliable beaconing scheme for vehicles due to the following three reasons. First, it is non-trivial to guarantee the road-safety demand for each vehicle with the limited available V2X bandwidth [4–7], especially under dense-vehicle conditions. On the one hand, if vehicles adopt aggressive beaconing rates, some vehicles may be sacrificed and have no required bandwidth to broadcast their moving status. On the other hand, if vehicles adopt moderate beaconing rates, the received moving status of neighboring vehicles may be out-of-date, resulting in delayed reactions to dangerous situations. In fact, vehicles in the moving usually have different danger levels, calling for distinct beaconing rates to enhance the road safety. Second, in vehicular environments, the lack of a global central unit makes it hard for optimal beaconing scheme achievement. Alternatively, vehicles have to negotiate in a fully *distributed* way, to access the available bandwidth in real time [8]. Third, as vehicles move fast and vehicular environments vary dramatically, the communication chance is limited with uncertainty. Therefore, the distributed beaconing scheme should minimize the communication overhead and react rapidly to the environment [9].

There have been two categories of studies on beacon congestion control, i.e., transmit power control (TPC) and transmit rate control (TRC). With predicting the vehicle distribution in advance, TPC schemes adjust transmission powers proactively to prevent future channel congestions, which are proactive solutions [10–14]. They are sensitive to estimation errors caused by biased transmission model and imprecise prediction model, which are common in vehicular environments with highly-dynamic mobilities. Therefore, they are unreliable in vehicular scenarios, which has been pointed out in the previous work[15]. In contrast, by controlling beaconing rates, TRC schemes react to congestion events that have already happened. However, within the bandwidth capacity constraint, current TRC schemes mainly focus on achieving the max-min fairness without considering safety-awareness. For instance, Linear MEssage Rate Integrated Control (LIMERIC) [16] and Periodically Updated Load Sensitive Adaptive Rate control (PULSAR) [17] consider equal fairness such that all vehicles within the congestion location takes the same level of beaconing rate adaptation. Such equal-fairness control schemes can achieve the maximum throughput gain but cannot satisfy road-safety demand of vehicles with different danger levels. With taking driving context into consideration, another two TRC schemes [18, 19] have been proposed recently, both of which formulated a network utility maximization problem to reply to the beacon rate adaptation. Particularly, each vehicle is associated with a utility function and the objective is to maximize the sum utilities of all vehicles. However, to quantify the utility function, the aggregated information is usually utilized, e.g., the accumulated velocities of one-hop neighbors and the accumulated relative distances with one-hop distances, which can hardly capture the road-safety demand of individual vehicle.

In this chapter, we propose a novel safety-aware *adaptive beacon control* scheme, named *ABC*, which can adjust vehicle beaconing rates adaptively in accordance with individual rear-end collision threat. By considering together the kinematic status of preceding-following vehicles and the beaconing rate, we first investigate a typical rear-end collision condition and define a *danger coefficient* ρ to capture individual rear-end collision threat, which can also implicitly indicate the beaconing bandwidth demand for the vehicle. The beaconing activities are considered under the context of the TDMA-based broadcast MAC protocol, which has been demonstrated with efficacy in supporting periodical broadcast communications. Then, given the total available channel capacity, we formulate a *distributed beacon rate adaptation* (DBRA) problem with all bandwidth requirements, which is proved to be NP-hard. To achieve adaptation result in real-time, we devise a heuristic greedy algorithm based on individual estimates of ρ. To be specific, with keeping the total required beaconing demand lower than the channel capacity, vehicles with higher estimated ρ will be assigned with higher beaconing rates, expecting that more frequent broadcasts are conducted by those vehicles to avoid rear-end collisions. In *ABC*, each vehicle estimates its own danger coefficient ρ and collects the information of ρ of neighboring vehicles through beacon exchanges, to be safety-aware. Based on the collected beaconing status, a *close-loop* control mechanism is then employed, in which each vehicle keeps identifying whether the channel is congested or not. Once a vehicle detects a channel congestion, it adopts the greedy algorithm to solve the DBRA problem locally and broadcasts the beacon rate adaptation results to neighboring vehicles, to mitigate the channel congestion. When a vehicle receives multiple inconsistent beacon adaptation results from its neighbors, it will conservatively adopt the lowest beaconing rate to avoid channel congestions. We theoretically analyze the performance of the devised heuristic algorithm, which is in close proximity to the optimal result.

For performance evaluation, based on Simulation of Urban MObility (SUMO)-generated vehicular traces, we implement the proposed *ABC* and conduct extensive simulations under various underlying road topologies and varied traffic densities. As a result, compared with two benchmark schemes, i.e., LIMERIC [16] and 802.11p [20], *ABC* can efficiently control the beaconing rates with complying with bandwidth constraint, and thus the rates of beacon transmission/reception collisions can be significantly reduced. Additionally, *ABC* can achieve beaconing reliability with satisfying road-safety demand of vehicles since beaconing rates are adjusted in accordance with crash threat of individual vehicles. The main contributions are threefold:

- We investigate the relationship between the beaconing rate and the according rear-end collision risk, based on which we define the danger coefficient ρ that captures the rear-end collision threat of each vehicle and describes the beaconing bandwidth demand with respect to such collision threats.

- Working under dynamic vehicular environments, we propose a fully-distributed adaptive beacon control scheme, named *ABC*, which can perform safety-aware beaconing rate adaptation for each vehicle. In the context of a TDMA-based broadcast MAC, we further formulate the DBRA problem and prove its NP-hardness. To solve the problem in real time, we devise a heuristic algorithm based on real-time estimated value ρ.
- We conduct theoretical performance analysis on the heuristic algorithm efficiency. In addition, we implement our proposed *ABC* under the SUMO-generated traces and conduct extensive simulations to demonstrate its efficacy.

We organize the rest of this chapter as follows. System model is presented in Sect. 7.2 and we investigate safety-aware beaconing rate adaptation in Sect. 7.3. Section 7.4 elaborates on the design of *ABC*. Performance analysis and performance evaluation are carried out in Sects. 7.5 and 7.6, respectively. We give a brief summary in Sect. 7.7.

7.2 System Model

For the system scenario, a set of moving vehicles and stationary RSUs are considered, and they communicate via DSRC radios. To achieve road-safety applications, they are required to broadcast beacons periodically, and the beaconing activities are considered in the context of TDMA-based broadcast MAC.

7.2.1 Dedicated Short Range Communications (DSRC)

To conduct wireless communications, all nodes[1] in the vehicular environment are equipped with a DSRC communication radio. The DSRC radio runs at the frequencies between 5.700 and 5.925 GHz, and supports channels with optional bandwidths of 10 and 20 MHz, which can provide peak rates of 27 and 54 Mbps, respectively[21, 22]. In DSRC, there are one CCH and multiple SCHs, where the CCH is used to deliver high-priority short messages (i.e., periodic and event-driven beacons) and control information (e.g., negotiations for SCHs usages) while the low-priority application data is delivered at SCHs [23]. One single DSRC radio can support both the CCH and SCHs by switching the channel every 50 ms. In this chapter, we focus on periodical beaconing activities at the CCH since they are the most important for road-safety application. In the network, we assume that radios have the identical communication capability, i.e., with the same communication range R. Therefore, we can denote the network with an undirected graph $G(V, E)$,

---

[1]In this chapter, nodes or vehicles are a broad term of the communicating vehicles and RSUs.

in which $V = \{1, 2, \ldots, n\}$ is the set of communicating nodes and E ($n \times n$ matrix) represents the link condition between any two nodes. Specifically, for two distinct nodes x and y, if they are within the communication range with each other, i.e., $d_{xy} \leq R$, then $E_{xy} = 1$, otherwise $E_{xy} = 0$. The one-hop neighbor set of node x is denoted by $N_{OHS}(x) = \{y \in V \mid y \neq x, d_{xy} \leq R\}$. To achieve beaconing reliability, beacons broadcasted by x should be successfully received by all one-hop neighboring vehicles in $N_{OHS}(x)$, otherwise potential dangerous situations may arise.

7.2.2 TDMA-Based Broadcast MAC

A TDMA-based broadcast MAC is considered to investigate upper-layer periodical beaconing activities, since the TDMA-based MAC has been demonstrated with efficacy in providing efficient broadcast communications [24–26]. The 802.11p MAC is not considered for two major reasons when supporting broadcast communications. First, as the 802.11p MAC works under a contention-based manner, it could result in unbounded delays when too many nodes contend for the channel simultaneously. Therefore, it is hard to satisfy the stringent delay requirement for periodical broadcasts when compared to the TDMA-based MAC that has a pre-determined frame and slot structure. Second, in the broadcast mode of 802.11p MAC, RTS/CTS packets are removed to facilitate real-time response, making the *hidden terminal problem* inundant [20]. It is worth noting that our proposed beaconing scheme can work independently from the underlying MAC protocol, which is readily applied in 802.11p by dynamically determining the back-off time, size of contention window, etc.

Time-Slotted Channel Medium In the TDMA-based MAC, we represent the medium resource of control channel by distinct time slots. As shown in Fig. 7.1, time is partitioned into consecutive frames, each containing a fixed number (denoted by S) of time slots, which is denoted by the set S. The time slot index is synchronized among vehicles via the GPS module (embedded in the DSRC radio). In the time-slotted channel, each vehicle applies for a unique time slot and broadcasts over it every frame. It can satisfy the stringent delay requirement for beacon-based road-safety applications by granting each vehicle a well-scheduled time slot.

Spatial Reuse TDMA Constraints The neighboring vehicles within the communication range of a vehicle, constitute its OHS. If the two OHSs overlap with each other, they can form a THS constituted by the union of these two OHSs. As shown in Fig. 7.1, the vehicle A locates in the OHS of both vehicle B and C, forming a THS. In each THS, vehicles can reach any other by at most two hops. Obviously, to avoid transmission collisions, vehicles in the same OHS should apply for different time slots. Additionally, vehicles in the same THS also have to access different time slots in order to overcome the hidden terminal problem. To be specific, without RTS/CTS mechanisms, hidden terminal problems arise in a THS when two

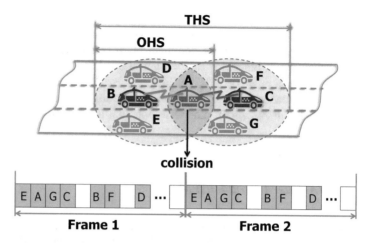

Fig. 7.1 Illustration of the TDMA-based broadcast MAC

vehicles cannot hear each other (e.g., locating in respective two OHSs) but decide to broadcast simultaneously. As shown in Fig. 7.1, as the vehicle B and C are out of the communication range of each other, they could perceive the channel to be free and it is very likely that they may broadcast in parallel, which can result in message collisions at the vehicle A. To cope with it, vehicles in the same THS should be allocated with distinct time slots. The THS (interference range) of vehicle x is denoted by $\mathcal{N}_{THS}(x)$, $\mathcal{N}_{THS}(x) = \mathcal{N}_{OHS}(x) \bigcup \{\mathcal{N}_{OHS}(y), \forall y \in \mathcal{N}_{OHS}(x)\}$. Figure 7.1 gives a valid time slot assignment as THS vehicles are allocated with different time slots.

Time Slot Access and Collision Detection To negotiate the time slot access in a fully distributed way, each vehicle (say vehicle x) includes the application data together with *frame information* in each beacon, denoted by $I(y)$ and $T(y)$ ($\forall y \in \mathcal{N}_{OHS}(x)$), representing the vehicle ID and the acquired time slot of y, respectively. Therefore, each vehicle can understand the THS time slot acquisition by listening to its OHS beacons. If the vehicle x wants to apply for a free time slot, it has to listen to the channel for S successive time slots to achieve the entire frame information, i.e., obtaining all $T(y)$ ($\forall y \in \mathcal{N}_{THS}(x)$). After that, it can choose a time slot from the vacant time slot set (i.e., $S - T(y)$ for $\forall y \in \mathcal{N}_{THS}(x)$) in a random way. Once a vehicle successfully acquires a time slot, it can use the same time slot in all subsequent frames unless a time slot collision happens. To detect the time slot collision in real time, at the end of each frame, vehicles can check the frame information received during previous S time slots. Particularly, for vehicle x, if all received beacons from y ($\forall y \in \mathcal{N}_{OHS}(x)$), indicating that $x \in \mathcal{N}_{OHS}(y)$ (i.e., the vehicle y successfully received the beacon from vehicle x), it means no concurrent transmissions happen at the slot $T(x)$ within its THS; otherwise, vehicle x may collide with another vehicle at the time slot $T(x)$. The vehicle has to release its original time slot and apply for a new one, if it detects a time slot collision.

Beaconing Starving Problem Under the TDMA-based broadcast MAC, we consider a normal beaconing case where a moderate transmission rate of 6 Mbps is adopted by the DSRC radio,[2] and each beacon takes up about 500 bytes [27], then about 0.67 ms is required to finish the data delivery. To support road-safety applications, each vehicle is required to broadcast every 100 ms, normally equaling the frame duration. Then, in each frame, the number of time slots is constrained to be no more than 150, where the physical layer overhead and guard period between time slots are not even taken into account. Besides, as found in the previous experimental study [2], the urban DSRC-enabled V2V communications are rather reliable within 300 m regardless of the fast speed and complex environments. It means that vehicles within the interference range, i.e., up to 1200 m, have to contend for the only 150 time slots. However, at urban intersections, on bidirectional 8-lane highways, or for traffic at rush hours, vehicles can heavily aggregate within small areas where the small number of time slots are far from enough to meet vehicle beaconing demands. This supply-demand imbalance problem is defined as the *beaconing starving problem*, which can result in channel congestions and thus deteriorate the beaconing performance. In this chapter, our goal is to cope with the beaconing starving problem while keeping vehicles being safety-aware.

7.3 Inferring Vehicle Hazardous Levels

In this section, before conducting the beacon rate adaptation, we delve into how the beaconing enhances the road safety. To this end, the kinematic status and beaconing rate are considered together to investigate the risk of encountering a rear-end collision. After that, a danger coefficient ρ is defined to capture the danger threat of each vehicle being in the collision. The rear-end collision situation is considered, since it is the most common type of vehicle crashes. According to the most recent report from the U.S. Department of Transportation [28], the rear-end crash alone accounts for more than 32% of all crashes. Based on our control framework, other types of crashes can be incrementally integrated to form a general coefficient, to be safety-aware for all types of crashes.

7.3.1 Danger Coefficient ρ

As shown in Fig. 7.2a, the following vehicle A (with a speed of V_f) moves after the preceding vehicle B (with a speed of V_p), and the following distance is denoted by d. With broadcasting communications, the vehicle B keeps reporting its moving

---

[2]The moderate transmission rate can provide better communication qualities considering the harsh wireless channel conditions, which has been verified with field experiments [2].

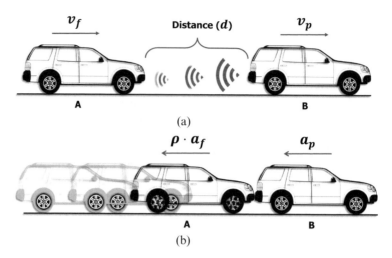

Fig. 7.2 Illustration of the kinematic status in rear-end collision situation. (**a**) Vehicle *A* moves after vehicle *B* normally. (**b**) A collision risk arises with the sudden deceleration from *B*

status to the vehicle *A* every T_{beacon} seconds (i.e., $\frac{1}{T_{\text{beacon}}}$ Hz). Then, the vehicle *A* can make a decision on acceleration or deceleration by sensing the moving status of vehicle *B* in real time. Figure 2.2b demonstrates a risk of encountering rear-end collision when the vehicle *B* taking a sudden brake with a_p m/s^2. The vehicle *A* can perceive the emergency rapidly and react to the situation with receiving the reporting beacon after a delay *T*, where $T = T_{\text{beacon}} + T_{\text{reaction}}$, and T_{reaction} is the reaction time for a driver (or a response time of the automatic controlling system for autonomous vehicles). The vehicle *A* has to brake a little or fully in order to avoid the collision with the preceding vehicle, which is determined by the current kinematic status of two vehicles.

Definition 7.1 (Danger Coefficient ρ) There are two vehicles *A* and *B* moving in the same lane, where *A* is the following vehicle while *B* is the preceding vehicle. If vehicle *B* decelerates suddenly at the maximum acceleration, after knowing the situation by receiving the beacon, vehicle *A* has to brake at ρ ($\rho \in (0, 1]$) times the maximum acceleration, in order to avoid a collision with *B*. Then, vehicle *B* is said to be dangerous with a coefficient ρ in terms of encountering a rear-end collision.

Therefore, as shown in Fig. 7.2, the kinematic relation of two vehicles satisfies

$$V_f(T_{\text{beacon}} + T_{\text{reaction}}) + (\frac{V_f^2}{2\rho a_f} - \frac{V_p^2}{2a_p}) = d, \tag{7.1}$$

where a_f is the maximum acceleration of vehicle *A*. Then, the danger coefficient ρ can be represented by

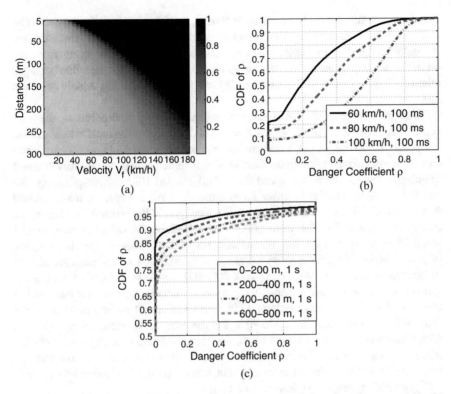

Fig. 7.3 Capturing danger threat through ρ. (**a**) The danger coefficient ρ vs. distance and V_f. (**b**) CDFs of ρ in different speed-limited lanes. (**c**) CDFs of ρ at intersection zones

$$\rho = \frac{V_f^2}{2a_f(d - V_f(T_{\text{beacon}} + T_{\text{reaction}}) + \frac{V_p^2}{2a_p})}. \tag{7.2}$$

By receiving beacons from neighboring vehicles constantly, each vehicle can update the moving status of the following vehicle and calculate the real-time danger coefficient ρ. The individual real-time danger coefficient can be shared among neighboring vehicles via beacon exchanges, which can be utilized to negotiate the beacon rate assignment in a fully distributed way.

7.3.2 Capturing Danger Threat

For each vehicle, the danger threat of encountering a potential collision can be indicated by the value of ρ. To be specific, as shown Fig. 7.3a, we plot the value of ρ under the function of velocity V_f and distance d, where V_p, a_f, a_p, T_{beacon}, and

$T_{reaction}$ is empirically set to be 60 km/h, 8 m/s^2, 8 m/s^2, 1 s, and 0.5 s, respectively. It shows that the underlying *driving context* can be well captured by the variation of coefficient ρ. For example, when the following vehicle is far away from the preceding vehicle or moves slowly, the value ρ is quite small, but the value becomes relatively large when the following vehicle is close to the preceding vehicle or moves fast.

To further understand how ρ varies in realistic driving environments, the ρ of vehicles are calculated in SUMO [29] every 100 ms, where the detailed simulation setup can be found in the performance evaluation section. Figure 7.3b shows cumulative distribution functions (CDFs) of ρ of vehicles (beaconing every 100 ms) in different lanes, each with a speed limit of 60, 80, and 100 km/h, respectively. We can observe that *vehicles in faster lanes normally have a larger ρ, which should be granted with higher beaconing rates*. For example, in the 60 km/h-limit lane, the median ρ (i.e., with the CDF value of 0.5) is about 0.2, but the value increases to 0.35 and 0.55 in the 80 and 100 km/h-limit lane, respectively. Figure 7.3c shows CDFs of ρ of vehicles (beaconing every 1000 ms) at intersection zones, each locating at the respective distance zone, i.e., 0–200, 200–400, 400–600, and 600–800 m, away from the center of intersection area. We can see that smaller ρ prevails at intersection zones, verified by two observations. First, when vehicles at intersection zones, the probability of $\rho = 0$ reaches up to 85%, but the probability decreases to as low as 10% (shown in Fig. 7.3b even the beaconing rate is ten times larger) when vehicles being far away from the intersection. Second, the probability of $\rho = 0$ decreases with the distance away from the intersection, and the proportion is about 85%, 79%, 73%, and 69% in respective intersection zones.

We can conclude that, *it is the intersection scenario that is urgently in need and suitable for applying the beacon congestion control, since there are more vehicles contending for the channel bandwidth in the dense-vehicle condition, and smaller values of ρ make it possible to adapt the beaconing rates*.

7.3.3 Safety-Aware Beacon Rate Adaptation

To enable upper-layer road-safety applications, vehicles usually can choose distinct time slots in a frame and broadcast with a rate of 1 beacon/frame. However, the beacon rate adaptation is required to suppress the congestion event when the beaconing starving problem happens. In this chapter, *we propose a scheme to dynamically adapt the beaconing rate for vehicles within a range* $[\alpha_{min}, \alpha_{max}]$. Taking the coefficient ρ as a driving factor, the following rule should be complied with during the beaconing resource allocation.

Rule 1 In the set of $\mathcal{N}_{\mathcal{THS}}(x) \cup x$, for two vehicles i and j, the beaconing rate of i should be greater than or equivalent to the beaconing rate of j, if $\rho_i \geq \rho_j$, i.e.,

$$\alpha_i \geq \alpha_j, \forall\{i, j | \rho_i \geq \rho_j, i, j \in \mathcal{N}_{\mathcal{THS}}(x) \cup x\}. \tag{7.3}$$

Additionally, in order to comply with the requirement of most road-safety applications [30], beaconing rates of vehicles should range from 1 to 10 Hz. Therefore, we set the $\alpha_{\min}$ and $\alpha_{\max}$ to be 0.1 and 1 beacon/frame,[3] respectively.

7.4 Design of ABC

7.4.1 Overview

In *ABC*, when the channel medium resource is sufficient, vehicles can usually broadcast at the maximum beaconing rate. Meanwhile, vehicles collect beaconing status of neighbors through sending/receiving beacons, to keep identifying whether the channel is congested. If one congestion event is identified by a vehicle, to suppress the congestion, a distributed beacon rate adaptation (DBRA) problem is then formulated at the vehicle to adapt beaconing rates for vehicles within the congestion. As the DBRA problem is proved to be NP-hard, the vehicle solves the problem locally by the devised heuristic algorithm. Finally, the vehicle will inform other vehicles (within the interference range) with the rescheduled results, based on which they can adapt beaconing rates accordingly to relieve the congestion. Considering the dynamic variation of ρ in the moving, to minimize the risk of encountering collision, the vehicle is allowed to increase the beaconing rate independently when its danger coefficient becomes larger than a threshold. In what follows, we elaborate on the key components of *ABC*: (1) online congestion detection; (2) distributed beacon rate adaptation; and (3) adaptation results informing. After that, we provide the adaptive beacon control approach in *ABC*.

7.4.2 Online Congestion Detection

Collecting Beaconing Status In the proposed scheme, to make vehicles being able to perceive the channel condition within the interference range (congested or not), vehicles are required to broadcast the application data together with the THS beaconing status. The individual beaconing status contains both the beaconing rate α and danger coefficient ρ. Specifically, for vehicles, they can continuously update their own danger coefficient based on newly received kinematic information, and then should include the up-to-date list information of THS (α, ρ) in each beacon. In doing so, with receiving beacons from the OHS neighbors, each vehicle can collect the real-time beaconing status of THS neighbors.

---

[3]The duration of each frame is normally set to be 100 ms.

Detecting Congestion Events To this end, by checking the up-to-date beaconing status of THS neighbors, each vehicle has the capability of detecting congestion events in real time.

Definition 7.2 (Beaconing Item Size) To model the beaconing resource usage, we define the *beaconing item size*, which is equivalent to α_i if the vehicle i has the beaconing rate of α_i beacon/frame ($\alpha_i \in (0, 1]$).

Definition 7.3 (Time Slot Space) To model the medium resource for beaconing activities, we define the *time slot space* for each time slot. One vacant time slot has the time slot space of 1, and for a non-vacant time slot, its time slot space is equivalent to 1 minus the sum of beaconing item sizes that the time slot is supporting.

Therefore, the vehicle x would detect a channel congestion event, if the beaconing rates of vehicles in its THS satisfy

$$\sum_{i=1}^{|\mathcal{N}_{\mathcal{THS}}(x)|+1} \alpha_i > S, \tag{7.4}$$

where $|\mathcal{N}_{\mathcal{THS}}(x)| + 1$ represents the number of vehicles in its THS including itself. At the end of each frame, the online congestion identification can be conducted based on received beacons.

7.4.3 Distributed Beacon Rate Adaptation (DBRA)

For those vehicles who have been involved in the congestion, they have to adapt beaconing rates to relieve the ongoing congestion. The following two constraints should be complied with.

Constraint 1 (Periodical Beaconing Rate) As the broadcast communication is periodical, the individual beaconing rate α_i within the THS of vehicle x satisfies

$$\alpha_i = \frac{1}{t}, t = 1, 2, 3, \ldots, 10, \forall i \in \mathcal{N}_{\mathcal{THS}}(x) \cup x, \tag{7.5}$$

i.e., beaconing every t frames while keeping silent during the other $t - 1$ frames.

Constraint 2 (Bandwidth Limitation) The total beaconing bandwidth is restrained within the channel capacity to avoid the channel congestion,[4] i.e.,

---

[4]It should be noted that, at the MAC layer, it is generally difficult to schedule the time slot usage without any wasting or colliding, especially for moving vehicles with diverse beaconing rates [31]. Therefore, it is better to leave some redundant time slots for scheduling to guarantee the MAC layer performance, which however is out of the scope of this chapter.

$$\sum_{i=1}^{|N_{THS}(x)|+1} \alpha_i \le S. \tag{7.6}$$

Safety-Weighted Network Utility Maximization The network utility usually depends on many variables such as throughput, delay, reliability, and among others. To build the network utility for road-safety applications, we incorporate the packet delay between communicating vehicles, which is the most important metric for delay-sensitive applications, together with the road-safety benefit, i.e., a multiplicative weight denoted by ρ. The packet delay can be impacted by many factors, including transmission data rate, medium access control, transmit power, and the quality of wireless link [32]. In our case, if one vehicle is assigned with a beacon rate $\alpha = \frac{1}{t}$, then the broadcast delay performance of the vehicle can be alternately represented by t, i.e., $1/\alpha$. In addition, as the network utility is generally relevant to the negative expected delay of each vehicle, the network utility contributed by the vehicle can be represented by α, since $\alpha \in (0, 1]$. The vector of beaconing rate assignments for THS vehicles is denoted by $\alpha = \{\alpha_i | i \in N_{THS}(x) \cup x\}$. To take the danger threat into consideration, we integrate the danger coefficient ρ into the utility function to represent the importance of road safety. Therefore, under the beaconing rates α, the safety-weighted network utility contributed by the vehicle i is

$$U_i(\alpha) = \rho_i \cdot \alpha_i, \forall i \in N_{THS}(x) \cup x. \tag{7.7}$$

We formulate the DBRA problem to maximize the sum utilities contributed by all vehicles involved in the congested THS:

$$\max \quad \sum_{i=1}^{|N_{THS}(x)|+1} \rho_i \cdot \alpha_i$$

$$s.t. \quad \alpha_i \in \{1/1, 1/2, 1/3, \ldots, 1/10\}, \tag{7.8}$$

$$\sum_{i=1}^{|N_{THS}(x)|+1} \alpha_i \le S.$$

If the objective function $\sum_{i=1}^{|N_{THS}(x)|+1} \rho_i \cdot \alpha_i$ can be maximized by adopting a specific α, the solution α will be able to comply with the requirement in **rule** 1. It can be easily proved by the contradiction. Specifically, if α is the optimal solution to the DBRA problem and (7.3) does not hold, there must exist $\alpha_i < \alpha_j$ while $\rho_i \ge \rho_j$. If the beaconing rates of i and j are exchanged with each other, a larger safety-weighted network utility to the DBRA problem can be achieved, which contradicts the maximizing property.

For each vehicle i, we can consider a class of beaconing rates, i.e., $C_i = \{1/1, 1/2, 1/3, \ldots, 1/10\}$, in which it has to choose one item j as its beaconing rate. We introduce a binary variable x_{ij} in our problem formulation, and it takes on value 1 if the item j is chosen otherwise it takes on 0. Then, we can equally formulate the DBRA problem as follows

$$max \quad \sum_{i=1}^{|\mathcal{N}_{\mathcal{THS}}(x)|+1} \sum_{j \in C_i} \rho_i \cdot \alpha_{ij} \cdot x_{ij}$$

$$s.t. \quad \sum_{i=1}^{|\mathcal{N}_{\mathcal{THS}}(x)|+1} \sum_{j \in C_i} \alpha_{ij} \cdot x_{ij} \leq S, \tag{7.9}$$

$$\sum_{j \in C_i} x_{ij} = 1, i = 1, \ldots, |\mathcal{N}_{\mathcal{THS}}(x)| + 1,$$

$$x_{ij} \in \{0, 1\}, i = 1, \ldots, |\mathcal{N}_{\mathcal{THS}}(x)| + 1, j \in C_i,$$

where α_{ij} represents that the vehicle i chooses the j-th beaconing rate in the class C.

We have the following theorem for the DBRA problem.

Theorem 7.1 *The DBRA problem is NP-hard.*

Proof To prove the NP-hardness of the DBRA problem, we can devise a polynomial reduction from a classic NP-hard problem to our problem. Specifically, we can devise from the *multiple-choice knapsack problem (MCKP)* [33], which is a variant of the ordinary 0–1 knapsack problem. In the **MCKP** problem, there are m mutually disjoint classes $U_1, U_2, \ldots, U_m$ of items, which need to be packed into a knapsack with a total capacity C, and each item $j \in U_i$ has a profit p_{ij} with a weight cost c_{ij}. The objective is to maximize the profit sum without exceeding the capacity C when choosing exactly one item from each class. Our DBRA problem as formulated in (7.9) is equivalent to the problem, concluding the proof.   □

Heuristic Algorithm for DBRA By adopting the dynamic programming (DP) algorithm, we can achieve the optimal result to the DBRA problem, but it requires a *pseudo-polynomial* time complexity, which is *unacceptable* for an online system. Particularly, when adopting the DP algorithm to solve the DBRA problem, the time complexity of $O(n \cdot S)$ is required, where n and S denotes the number of input vehicles and total capacity of all time slots, respectively. Intuitively, the complexity $O(n \cdot S)$ looks like a polynomial time in terms of input values n and S. However, for the input size, to represent the value of S, 2^L bits are required, and then the time complexity becomes $O(n \cdot 2^L)$, which is an exponential time rather than a polynomial time in terms of input size. Therefore, to achieve real-time decisions in ABC, we devise a heuristic greedy algorithm to conduct beaconing rate adaptation. More specifically, all vehicles $\mathcal{N}_{\mathcal{THS}}(x) \cup x$ are first granted with the minimum beaconing rate α_{min}. To assign the remaining medium resource, vehicles are then sorted by the danger coefficient ρ in descending order; for the vehicle with the largest ρ, it will be allocated with more beaconing resource until reaching α_{max}. The procedure repeats until all the medium resources are used up. Algorithm 1 presents the pseudocode of the greedy algorithm. It should be pointed out that, the vehicle danger coefficient ρ could vary constantly with fast moving. As a result, after a period of time, the previous adaptation results may cause unfairness. For instance, the vehicle danger coefficient ρ may become larger when it leaves the intersection, which could result in potential dangers. In the design of *ABC*, vehicles are allowed to increase beaconing rates independently to avoid such type of potential dangers, once their danger coefficients reach up to a pre-defined threshold.[5]

Algorithm 1 Heuristic Algorithm for DBRA at the vehicle x

Input: S, α_{min}, α_{max} and $N_{THS}(x) \cup x$
Output: α, i.e., $\{\alpha_i | i \in N_{THS}(x) \cup x\}$
1: Initialize: $\alpha = 0$
2: **for** $i \in N_{THS}(x) \cup x$ **do**
3:     $\alpha_i = \alpha_{min}$
4: **end for**
5: Left_capacity$= S - \alpha_{min} \cdot |N_{THS}(x) \cup x|$
6: $Sort(N_{THS}(x) \cup x)$ with $\downarrow \rho_i$
7: **while** Left_capacity> 0 **do**
8:     $id \leftarrow (N_{THS}(x) \cup x)[0]$
9:     **if** Left_capacity $\geq \alpha_{max} - \alpha_{min}$ **then**
10:         $\alpha_{id} = \alpha_{max}$
11:         Left_capacity$- = \alpha_{max} - \alpha_{min}$
12:         $N_{THS}(x) \cup x = N_{THS}(x) \cup x - \{id\}$
13:     **else**
14:         $\alpha_{id} = \alpha_{min} +$Left_capacity
15:         Left_capacity$= 0$
16:         **break**
17:     **end if**
18: **end while**
19: **return** α

7.4.4 Adaptation Result Informing

At the congested location, after the vehicle (say vehicle A) solves the DBRA problem, it has to inform other THS vehicles with the adaptation results. Particularly, the vehicle A first compares the adaptation results to its THS beaconing status. For those vehicles, whose current beaconing rates are larger than the adaptation result, A will inform them via the next beacon transmission with including the informing information that contains a list of (vehicle ID, assigned beaconing rate). After a neighboring vehicle (say vehicle B) receives the informing beacon, it will compare the individual beaconing rate with the allocated result, and reduce the beaconing rate accordingly if its beaconing rate is larger. In addition, the vehicle B also compares

---

[5]The threshold setting does not affect how *ABC* works, and thus is conservatively set to be 0.5 in the work.

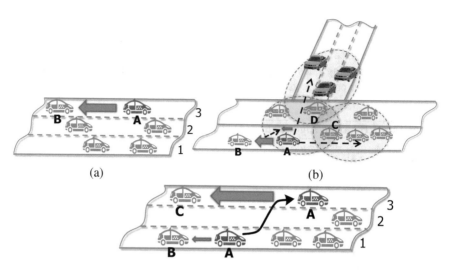

Fig. 7.4 Examples of adaptive beacon control. (**a**) When the resource is sufficient. (**b**) Encountering a congestion event. (**c**) Road safety demands increase

the informing results with the OHS (who may be out of the communication range of vehicle A) beaconing status, and includes the informing information in the next beacon to inform those vehicles who are required to adapt beaconing rates. With two major reasons, we do not adjust the beaconing rates for those vehicles whose current beaconing rates are smaller than the allocated results. First, although they have little impact on the current congested THS, they may trigger other THS congestions after improving their beaconing rates. Second, as the number of THS vehicles is large, it would result in significant communication overhead with informing all vehicles, and more importantly, it can prolong the convergence to the stable state. Note that, the adaptation results have to be disseminated within the full congested THS, in order to reach the stable state. For the convergence performance of ABC, we consider the worst case that the vehicle A is assigned with the minimum beaconing rate. Then, one second (beaconing every 1 s) is required to make the informing beacon reach the full OHS of vehicle A. Assuming one of the one-hop neighbor is also assigned with the minimum beaconing rate, another one second may be required to cover all THS neighbors. Nevertheless, the congestion control can converge to the stable state within at most 2 s.

7.4.5 Adaptive Beacon Control Approach

Figure 7.4 illustrates how the proposed ABC scheme works. Specifically, as shown in Fig. 7.4a, when the vehicle A moves in the fast lane where the current traffic density is light, it can broadcast at the maximum beaconing rate to minimize the risk of being collided. Note that, in figures, we represent the size of beaconing rate by the width of arrow. However, as shown in Fig. 7.4b, multiple sets of vehicles from different road segments can merge at the intersection, and when the vehicle A approaches the intersection, it is very likely to meet a congested channel condition. In addition, the red traffic light can make vehicles slow down until they completely stop, which also aggravates the vehicle density. After the vehicle A identifies the congested event, it will solve the DBRA problem locally and broadcast out the adaptation results (shown by two long dashed arrows). Its one-hop neighbors C and D will also broadcast the required adjustments to cover all vehicles within the congestion location. The short dashed arrow indicates that the vehicle A adapts from the original larger beacon rate (represented by the large blue arrow) to a new smaller beacon rate (represented by the small blue arrow). It is possible that one vehicle may receive multiple inconsistent beaconing adaptation results from different congestion locations. To relieve the congestion events, the vehicle will conservatively adopt the lowest beaconing rate under this case. Figure 7.4c shows another example, where the vehicle A initially moves in the slow lane and leaves the intersection. It then changes to the fast lane and speeds up, and therefore the danger threat ρ goes up. Its beaconing rate will be increased accordingly due to the increased road-safety demand. The pseudocode of ABC scheme is given in Algorithm 2, which is executed at each vehicle to conduct adaptive beacon control.

7.5 Performance Analysis

In this section, we theoretically analyze the efficiency of heuristic algorithm by demonstrating its slight performance gap to the optimal result.

To begin with, we first sort vehicles by the danger coefficient in descending order, i.e., $\rho_i \geq \rho_j$ for $i < j$. Then, if $\sum_{j=1}^{i} \alpha_{\max} \leq S - (|\mathcal{N}_{THS}(x)| + 1 - i) \cdot \alpha_{\min}$, we can assign the i-th vehicle with the maximum beacon rate $\alpha_{\max}$, since the remaining $(|\mathcal{N}_{THS}(x)| + 1 - i)$ vehicles can be guaranteed with the minimum beacon rate. The first vehicle that is not assigned with the maximum beaconing rate is defined as the breaking item, and we can represent the index of the breaking item b as

$$b = \arg \min_i \sum_{j=1}^{i} \alpha_{\max} > S - (|\mathcal{N}_{THS}(x)| + 1 - i) \cdot \alpha_{\min}. \qquad (7.10)$$

Therefore, we can express the adaptation results (calculated by the heuristic algorithm) as

$$\alpha_i = \begin{cases} \alpha_{\max} & i < b; \\ S - (|\mathcal{N}_{THS}(x)| + 1 - b) \cdot \alpha_{\min} - \sum_{j=1}^{b-1} \alpha_j & i = b; \\ \alpha_{\min} & i > b. \end{cases} \qquad (7.11)$$

Algorithm 2 Algorithm for ABC at the vehicle x

Input: S, ρ_threshold, α_{min} and α_{max}
Output: beaconing rate α_x
 1: Initialize: channel_state = free, $\rho_x = 1$, $\mathcal{N}_{THS}(x)=\{\}$, $\mathcal{N}_{OHS}(x) = \{\}$, $\alpha_x = \alpha_{max}$
 2: **while** at the end of each frame **do**
 3:     Update $\mathcal{N}_{THS}(x)$ and $\mathcal{N}_{OHS}(x)$
 4:     Calculate ρ_x
 5:     total_load = 0
 6:     **for** $i \in \mathcal{N}_{THS}(x)$ **do**
 7:         total_load = total_load + α_i
 8:     **end for**
 9:     **if** total_load + $\alpha_x > S$ **then**
10:         channel_state = congested
11:         $\boldsymbol{\alpha} = DBRA(S, \alpha_{min}, \alpha_{max}, \mathcal{N}_{THS}(x) \cup x)$
12:         $\alpha_x = \boldsymbol{\alpha}[x]$
13:         Inform results
14:     **else**
15:         channel_state = free
16:         **if** $\rho_x > \rho$_threshold **then**
17:             $\alpha_x = \alpha_{max}$
18:         **end if**
19:     **end if**
20:     **if** Receive informing result list L_y **then**
21:         **for** $id \in L_y$ **do**
22:             **if** $id == x$ and $L_y[id] < \alpha_x$ **then**
23:                 $\alpha_x = L_y[id]$
24:             **end if**
25:         **end for**
26:         Reinform results
27:     **end if**
28: **end while**

For convenience, the S time slots are divided into two groups, i.e., $S_1 = \{1, 2, \ldots, b - 1\}$ and $S_2 = \{b, b + 1, \ldots, S\}$. The subgroup S_1 is used to support the preceding $b - 1$ vehicles with the maximum beaconing rate, while the subgroup S_2 is used to guarantee the minimum beaconing rate for the rest of vehicles.

Lemma 7.1 *By adopting our heuristic algorithm, the safety-weighted network utility contributed by the subgroup S_1 can be as the same as the optimal result.*

Proof It can be proved with two steps. First, for vehicles $\{1, 2, \ldots, b - 1\}$, we consider they are served by the time slots in S_1 and assigned with the maximum beaconing rate. For the subgroup S_1, the network utility has been maximized since the time slots are fully utilized by those vehicles that have the maximum danger coefficients. Second, without assigning time slots in S_1 to those vehicles, no larger network utility can be achieved. It can be easily proved by contradiction. If a time slot in S_1 is assigned to a group of vehicles in the set $\{b, b + 1, \ldots, S\}$ and a larger network utility can be achieved, then there exists $i > j$ while $\rho_i > \rho_j$, which contradicts the descending order property. $\qquad\square$

For vehicles $\{b, b + 1, \ldots, |\mathcal{N}_{\mathcal{THS}}(x)| + 1\}$, to guarantee their minimum beaconing rates, a total $(|\mathcal{N}_{\mathcal{THS}}(x)| + 1 - (b - 1)) \cdot \alpha_{min}$ time slot spaces should be kept for time slots in subgroup $\mathcal{S}_2$.

Definition 7.4 (Uncertain Space) In subgroup $\mathcal{S}_2$, there is an uncertain space, equaling $S - (b - 1) - (|\mathcal{N}_{\mathcal{THS}}(x)| + 1 - (b - 1)) \cdot \alpha_{min}$. The uncertain space could support one large-size vehicle or a combination of small-size vehicles (in terms of beacon rate), which can lead to the knapsack complexity.

Lemma 7.2 *The uncertain space cannot exceed the size* $\alpha_{max} - \alpha_{min}$.

Proof It can be easily proved since if the uncertain space is larger than $\alpha_{max} - \alpha_{min}$, then another vehicle (i.e., b-th vehicle) can be allocated with the maximum beaconing rate, which contradicts the breaking item property of the b-th vehicle. □

To this end, the performance gap between the devised heuristic algorithm and the optimal result can only happen at the usage of uncertain space. Particularly, the heuristic algorithm may fail to make full use of it with assigning to only one vehicle that has a large danger coefficient, but the DP algorithm can fully utilize it by dynamic combination. As the size of uncertain space is small, the devised heuristic algorithm can achieve the near-optimal performance in general. To further demonstrate it, we implement both algorithms and carry out their numerical results. To be specific, the number of time slots S is set to be 150, and the number of vehicles $|\mathcal{N}_{\mathcal{THS}}(x)| + 1$ (input size) is varied from 50 to 450 with a step of 50. In addition, the danger threat of each vehicle ρ_i is randomly chosen from 0 to 1. The value of safety-weighted network utility in Eq. (7.9) is calculated, and the execution time for two algorithms are logged with a normal laptop.

The average results are shown in Fig. 7.5 after running the simulation for 20 rounds. As shown in Fig. 7.5a, compared with the DP algorithm, it can be seen that our devised heuristic algorithm can achieve quite analogous performance with it in terms of safety-weighted network utility. For instance, when with 200 vehicles, the network utility achieved by the heuristic and DP algorithm is about 93.90 and 93.91, respectively, which is a negligible performance gap. However, there is a huge gap between two algorithms in terms of the running time, shown in Fig. 7.5b. Particularly, compared with the DP algorithm with a time complexity of $O(n \cdot 2^L)$, our devised heuristic algorithm has a polynomial time complexity of $O(n log n)$ to obtain the result. For instance, no matter how many vehicles need to be scheduled, the required running time never exceeds 0.5 ms for the heuristic algorithm, which can ably meet the real-time requirement. On the contrary, when the number of vehicles is only 10, the required running time reaches about 130 s for the DP algorithm, and when the input size is increased to 450, the time can exceed even above 3200 s.

Fig. 7.5 Efficiency of the
heuristic algorithm. (**a**)
Network utility vs. input size.
(**b**) Running time vs. input
size

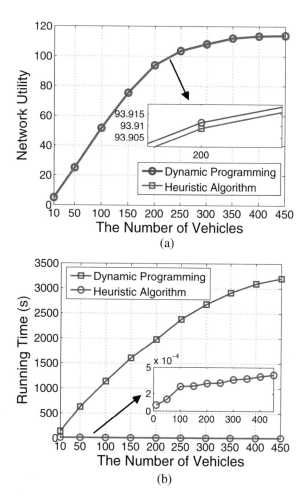

(a)

(b)

7.6 Performance Evaluation

7.6.1 Methodology

Simulation Setup To build real driving scenarios, the SUMO is adopted to create
an intermodal traffic system, which includes road topologies, traffic lights, moving
vehicles, etc. The main simulation parameters are summarized in Table 7.1. To
be specific, we create a typical urban road topology with four bidirectional 6-
lane road segments merging at the center and forming an intersection. Each road
lasts 5 km long, and three lanes in each direction are set with a speed limit of
60, 80, and 100 km/h, respectively. For each road segment, traffic lights are set
at the intersection with the green light, yellow light, and red light being 20 s, 3 s,
and 20 s, respectively. To emulate normal traffic conditions, vehicles are generated

Table 7.1 Simulation parameters

Parameters	Urban
Road length	5 km
Number of road segments	4
Number of lanes on each road	6
Speed limit in lanes (in km/h)	[60, 100]
Maximum speed of vehicles (in km/h)	[80, 240]
Acceleration of vehicles (in m/s^2)	[1.0, 5.0]
Deceleration of vehicles (in m/s^2)	[3.0, 10.0]
Transmission range	300 m
Frame duration	100 ms
Number of slots (per frame)	150
Loaded vehicles	800–1080
Simulation time	1000 frames

at the open end of each road segment with a rate of 10 vehicles/lane/min. For vehicle roadworthiness, the maximum velocity is ranged from 80 to 240 km/h, acceleration capability is ranged from 1 to 5 m/s^2, and deceleration capability is ranged from 3 to 10 m/s^2, which are randomly determined within the range when vehicles are generated. Vehicles are driven under the LC2013 lane-changing and Krauss car-following model [29], where normal lane-change and overtaking events are conducted when necessary. After a vehicle is generated, it will randomly choose a destination and will disappear from the system when reaching the destination. Figure 7.6 shows the snapshots of the simulated scenarios, in which different vehicle colors represent distinct roadworthiness sets.

We implement our proposed *ABC* scheme based on the generated SUMO trace, where 1000 frames of trace are utilized. As we study network activities, we consider all transmissions within the communication range are successful unless time slot usage collisions happen. We set the transmission range R to be 300 m according to the DSRC experiment study [2], which verifies that V2V communication can be rather reliable within the distance. In addition, in order to comply with the 100 ms delay requirements for most road-safety applications [30], each frame is set to 100 ms with 150 time slots.

Benchmarks We consider the following two benchmarks for performance comparison.

- **Conventional 802.11p [20]:** In broadcast mode of IEEE 802.11p, all vehicles broadcast with a rate of 1 beacon/frame, and there is no any congestion control mechanism.
- **LIMERIC [16]:** In LIMERIC, once a congestion event is identified, vehicles within the interference range will adapt to the same beaconing rate under the medium resource limit. Note that, this scheme normally works with different system models like our scheme, and we just adopt its congestion control idea as a benchmark here, i.e., treating all vehicles equally when a congestion event is identified.

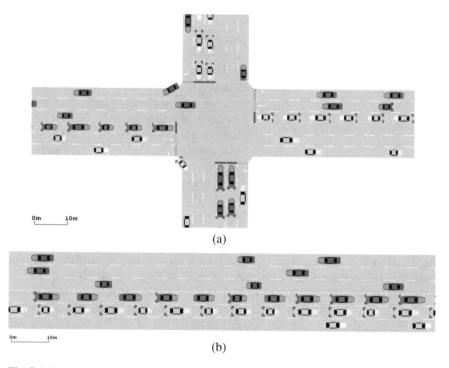

(a)

(b)

Fig. 7.6 Snapshots of the simulated scenario. (**a**) At intersection. (**b**) On multi-lane road

Benchmarks are implemented by Python in the same way as our *ABC* scheme does.

Performance Metrics We define the following five metrics to evaluate the performance.

- **Rate of beacon transmissions**: refers to the average number of transmitted beacons per frame.
- **Efficiency ratio of transmissions**: refers to the ratio of the number of successful transmissions to the total number of transmissions. When a vehicle transmits a beacon at a time slot and no concurrent transmission happens within its interference range, then the transmission is counted as a successful transmission.
- **Rate of beacon receptions**: refers to the average number of successfully received beacons per frame.
- **Reception coverage ratio**: refers to the number of successful receptions at a vehicle to the total number of transmissions by its OHS neighbors in a frame, measuring whether the vehicle can successfully receive all beacons from neighboring vehicles.
- **Rate of reception collisions**: refers to the average number of reception collisions per frame. Particularly, if a receiver receives more than one beacon at a single time slot, then the number of concurrent receptions are counted as the number of reception collisions.

7.6.2 Performance Comparison

To examine the system performance, we first check the CDF results of all metrics, where the results of all vehicles are aggregated.

Efficiency of Beacon Rate Control Figure 7.7a shows the number of accumulated congestions in the system, where we can observe that with beacon congestion control, *ABC* and LIMERIC can successfully suppress the congestion events. However, in 802.11p, as there is no congestion control, the congestion events can flood the system, with 20,000 times at the 150-th frame. To figure out how the beaconing activities are controlled by the schemes of *ABC* and LIMERIC, we plot the CDF results of rate of beacon transmissions in Fig. 7.7b, where we can observe that the beacon transmission rates are reduced effectively in both schemes. For instance, in the schemes of *ABC* and LIMERIC, the median beacon transmission rate is about 825 and 800 transmissions/frame, respectively, but in 802.11p, the rate can be as high as 950 transmissions/frame. In addition, in the schemes of *ABC* and LIMERIC, the maximum beacon transmission rate is less than 900 transmissions/frame, but the value can reach up to 1080 in 802.11p.

With suppressing congestion events properly, the transmission reliability can be significantly enhanced. Taking the metric of efficiency transmission ratio as an example (shown in Fig. 7.7c), the median ratio is about 0.56 by adopting 802.11p, but the value can be enhanced to 0.75 and 0.65 when adopting *ABC* and LIMERIC, respectively. Moreover, in *ABC*, more than 98% ratios are larger than 0.7, but the proportion drops down to 10% and 0 when adopting LIMERIC and 802.11p, respectively. It demonstrates that after a beacon is broadcasted out, the probability of successfully receiving it by all neighboring vehicles can reach about 70% with adopting *ABC*. To further indicate the important role of beacon congestion control, we plot the CDFs of reception collision rates in Fig. 7.7f, where we can observe that the median reception collision rate can be reduced to 5000 collisions/frame in *ABC*, but the value reaches over 12,000 and 25,000 when adopting the LIMERIC and 802.11p, respectively. Additionally, with adopting *ABC*, all reception collision rates are smaller than 10,000 collisions/frame, while more than 60 and 99% rates are larger than it when adopting the LIMERIC and 802.11p, respectively.

Slight Degradation in Rx Throughput We plot the CDFs of beacon reception rates in Fig. 7.7d, where it can be seen that 802.11p can sometimes achieve the largest beacon reception rates when the vehicle density becomes heavy.[6] For instance, with adopting *ABC* and LIMERIC, the median reception rate is about 29,000 and 26,000, which can be improved slightly to 29,500 in 802.11p. This phenomenon is reasonable since transmission rates are reduced in both *ABC* and LIMERIC schemes. However, although massive reception collisions would happen when adopting 802.11p, it is still possible to maintain the Rx throughput with the

---

[6]The vehicle density can be indirectly represented by the rate of reception.

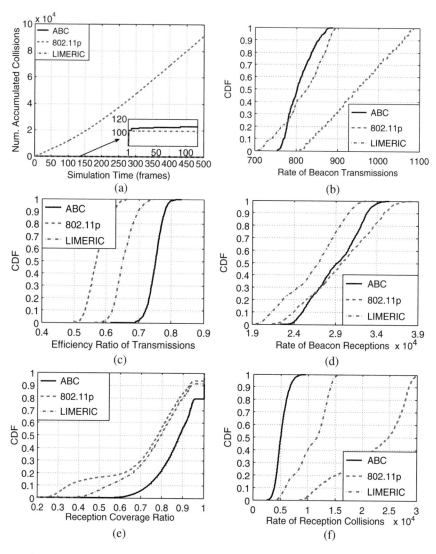

Fig. 7.7 Performance comparison. (**a**) Number of accumulated congestions. (**b**) CDFs of rate of beacon transmissions. (**c**) CDFs of efficiency ratio of transmissions. (**d**) CDFs of rate of beacon receptions. (**e**) CDFs of reception coverage ratio. (**f**) CDFs of rate of reception collisions

maximum transmission rate of all vehicles. It should be pointed out that a slight degradation of Rx throughput is *acceptable* in the vehicular network, since the goal of system is to provide reliable broadcast communications for road safety enhancement rather than maximizing the network throughput while sacrificing their road safety. On the other hand, as shown in Fig. 7.7e, when adopting 802.11p and LIMERIC, the probability of the reception coverage ratio being larger than 0.8 is just about 45 and 50%, but when adopting our proposed *ABC*, the probability can be enhanced to 80%. It means that vehicles are better aware of the nearby environment since they can receive beacons from more neighboring vehicles successfully.

7.6.3   Working Robustly Under Dynamic Road Traffic

To investigate the performance of *ABC* under different traffic conditions, we divide vehicles into two groups according to the number of THS vehicles, i.e., $|\mathcal{N}_{THS}(x)| + 1 \geq 150$ (under heavy traffic (HT)) and $|\mathcal{N}_{THS}(x)| + 1 < 150$ (under light traffic (LT)). We plot the system performance under different conditions in Fig. 7.8, where the solid lines and dashed lines are the results under the LT and HT conditions, respectively. As shown in Fig. 7.8a and b, we can observe that *ABC* can achieve the supreme congestion control performance under HT conditions. For instance, the median efficiency transmission ratio is only 0.25 in 802.11p, which can be enhanced to 0.61 when adopting *ABC*. Moreover, the median reception

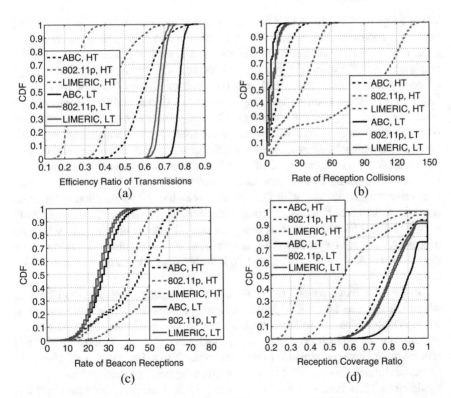

Fig. 7.8 Performance of vehicles under LT and HT conditions. (**a**) CDFs of efficiency ratio of transmissions. (**b**) CDFs of rate of reception collisions. (**c**) CDFs of rate of beacon receptions. (**d**) CDFs of reception coverage ratio

collision rate reaches over 100 collisions/frame/vehicle in 802.11p, which can be significantly reduced to 15 by adopting *ABC*. On the other hand, in both figures, we can observe that *ABC* can achieve the best performance in LT conditions, i.e., with the highest efficiency transmission ratio and the lowest reception collision rate. However, for the CDFs of beacon reception rates shown in Fig. 7.8c, we observe that *ABC* can outperform 802.11p under LT conditions but perform a little worse under HT conditions. It can be concluded that the higher efficiency transmission ratio of *ABC* can ensure a better RX throughput under LT conditions, but when under HT conditions, the large number of vehicles with maximum beaconing rates in 802.11p could result in a larger Rx throughput. Nevertheless, as shown in Fig. 7.8d, the reception coverage ratio can be significantly enhanced in *ABC* under both density conditions. For example, when adopting *ABC*, LIMERIC, and 802.11p, the median reception coverage ratio is about 0.79 0.58, and 0.38 respectively.

7.6.4 Safety-Aware Beacon Rate Adaptation

When adopting *ABC*, the CDFs of average beaconing rate of vehicles in different lanes are shown in Fig. 7.9a, where we can observe that with possible higher danger coefficients, vehicles in fast lanes are assigned with higher beacon rates. For instance, in the respective 60, 80, and 100 km/s-limit lane, the median average beaconing rate is about 0.93, 0.94, and 0.98 beacons/frame. Besides, we plot the CDFs of average beaconing rate in Fig. 7.9b when vehicles at different intersection zones, i.e., with respective 0–200, 200–400, 400–600, and 600–800 m away from the intersection center area. We can make the following two statements. First, the average beaconing rate can increase with the distance away from the intersection. For example, when in the range of 0–200 m, the median average beaconing rate is about 0.35, which can be increased to 0.68, 0.7, and 0.8 when within other respective ranges. Second, with the crowded vehicles, the intersection is the main scenario to conduct the beacon congestion control. Particularly, compared with results in Fig. 7.9a when vehicles being far away from the intersection, where all beaconing rates are larger than 0.88, the percentage of beaconing rates being larger than the value is about 0, 0, 15% and 30% within respective ranges at intersection areas.

On the other hand, we plot the CDFs of average danger coefficients in Fig. 7.9c, where we can see that 802.11p can achieve the smallest danger coefficients. It is reasonable since all vehicles broadcast with the maximum beaconing rates. In addition, we can observe that with a negligible gap, our proposed *ABC* can achieve quite close performance to 802.11p in terms of safety-awareness. However, as demonstrated above, 802.11p suffers from serious channel congestion with significant packet loss, which can result in potential dangers. Moreover, we can observe an obvious performance gap between *ABC* and LIMERIC, where they achieve the median average danger coefficient of 0.299 and 0.302, respectively. It happens because beaconing rates are assigned according to the underlying

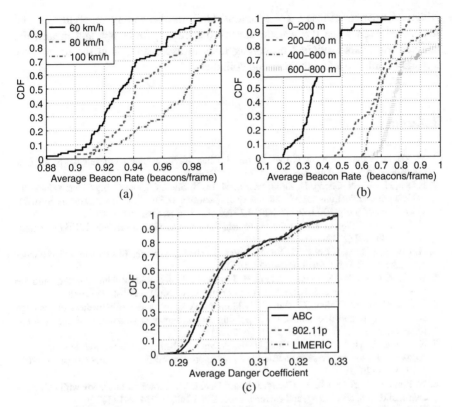

Fig. 7.9 Safety-aware beacon rate adaptation results. (**a**) Average beacon rate in different lanes. (**b**) Average beacon rate at intersection zones. (**c**) CDFs of average danger coefficients

driving contexts in *ABC*, but in LIMERIC, beaconing rates are equally assigned. To summarize, under congested channel conditions, *ABC* can control beaconing activities to avoid massive message collisions, and meanwhile can conduct safety-aware beaconing rate adaptation for vehicles.

7.7 Summary

In this chapter, we have proposed a distributed adaptive beacon control scheme, named *ABC*, to conduct safety-aware beacon rate adaptation for vehicles under highly-dynamic vehicular environments. Three novel techniques have been integrated in *ABC*: (1) online congestion detection; (2) distributed beacon rate adaptation; and (3) adaptation results informing. Performance analysis on the efficiency of the devised algorithm has been provided. Also, we have implemented our proposed *ABC* scheme under SUMO-generated traces and conducted extensive simulations to demonstrate its efficacy. By adopting *ABC*, the beaconing reliability can be guaranteed even when the vehicle density becomes quite heavy, and meanwhile vehicles can adapt sufficient beaconing rates in accordance with the road-safety demand in order to avoid the rear-end collisions.

References

1. W. Xu, H. Zhou, N. Cheng, F. Lyu, W. Shi, J. Chen, X. Shen, Internet of vehicles in big data era. IEEE/CAA J. Autom. Sinica **5**(1), 19–35 (2018)
2. F. Lyu, H. Zhu, N. Cheng, H. Zhou, W. Xu, M. Li, X. Shen, Characterizing Urban vehicle-to-vehicle communications for reliable safety applications. IEEE Trans. Intell. Transp. Syst. **21**, 1–17. https://doi.org/10.1109/TITS.2019.2920813. Early Access, Jun. 2019
3. X. Cheng, R. Zhang, L. Yang, Wireless toward the era of intelligent vehicles. IEEE Int. Things J. **6**(1), 188–202 (2019)
4. N. Cheng, F. Lyu, J. Chen, W. Xu, H. Zhou, S. Zhang, X. Shen, Big data driven vehicular networks. IEEE Netw. **32**(6), 160–167 (2018)
5. D. Yu, Y. Zou, J. Yu, X. Cheng, Q. Hua, H. Jin, F.C.M. Lau, Stable local broadcast in multihop wireless networks under SINR. IEEE-ACM Trans. Netw. **26**(3), 1278–1291 (2018)
6. S.M.O. Gani, Y.P. Fallah, G. Bansal, T. Shimizu, A study of the effectiveness of message content, length, and rate control for improving map accuracy in automated driving systems. IEEE Trans. Intell. Transp. Syst. **20**, 1–16 (2018)
7. N. Cheng, F. Lyu, W. Quan, C. Zhou, H. He, W. Shi, X. Shen, Space/aerial-assisted computing offloading for IoT applications: a learning-based approach. IEEE J. Sel. Areas Commun. **37**(5), 1117–1129 (2019)
8. W. Zhuang, Q. Ye, F. Lyu, N. Cheng, J. Ren, SDN/NFV-empowered future iov with enhanced communication, computing, and caching. Proc. IEEE **108**(2), 274–291 (2020)
9. H. Zhou, N. Cheng, Q. Yu, X. Shen, D. Shan, F. Bai, Toward multi-radio vehicular data piping for dynamic DSRC/TVWS spectrum sharing. IEEE J. Sel. Areas Commun. **34**(10), 2575–2588 (2016)
10. S.A.A. Shah, E. Ahmed, J.J.P.C. Rodrigues, I. Ali, R.M. Noor, Shapely value perspective on adapting transmit power for periodic vehicular communications. IEEE Trans. Intell. Transp. Syst. **19**(3), 977–986 (2018)
11. M. Torrent-Moreno, J. Mittag, P. Santi, H. Hartenstein, Vehicle-to-vehicle communication: fair transmit power control for safety-critical information. IEEE Trans. Veh. Technol. **58**(7), 3684–3703 (2009)
12. F. Goudarzi, H. Asgari, Non-cooperative beacon power control for VANETs. IEEE Trans. Intell. Transp. Syst. **20**, 1–6 (2018)
13. E. Egea-Lopez, P. Pavon-Mariuo, Fair congestion control in vehicular networks with beaconing rate adaptation at multiple transmit powers. IEEE Trans. Veh. Technol. **65**(6), 3888–3903 (2016)
14. J. Mittag, F. Schmidt-Eisenlohr, M. Killat, J. Härri, H. Hartenstein, Analysis and design of effective and low-overhead transmission power control for VANETs, in *VANET '08: Proceedings of the fifth ACM International Workshop on VehiculAr Inter-NETworking* (2008), pp. 39–48
15. M. Sepulcre, J. Mittag, P. Santi, H. Hartenstein, J. Gozalvez, Congestion and awareness control in cooperative vehicular systems. Proc. IEEE **99**(7), 1260–1279 (2011)
16. G. Bansal, J.B. Kenney, C.E. Rohrs, LIMERIC: a linear adaptive message rate algorithm for DSRC congestion control. IEEE Trans. Veh. Technol. **62**(9), 4182–4197 (2013)
17. T. Tielert, D. Jiang, Q. Chen, L. Delgrossi, H. Hartenstein, Design methodology and evaluation of rate adaptation based congestion control for vehicle safety communications, in *2011 IEEE Vehicular Networking Conference (VNC)* (2011), pp. 116–123

18. E. Egea-Lopez, P. Pavon-Mariuo, Distributed and fair beaconing rate adaptation for congestion control in vehicular networks. IEEE Trans. Mobile Comput. **15**(12), 3028–3041 (2016)
19. L. Zhang, S. Valaee, Congestion control for vehicular networks with safety-awareness. IEEE/ACM Trans. Netw. **24**(6), 3290–3299 (2016)
20. ASTM, Standard Specification for Telecommunications and Information Exchange Between Roadside and Vehicle Systems — 5 GHz Band Dedicated Short Range Communications (DSRC) Medium Access Control (MAC) and Physical Layer (PHY) Specifications. https:www.astm.org/Standards/E2213.htm
21. H. Zhou, W. Xu, J. Chen, W. Wang, Evolutionary V2X technologies toward the internet of vehicles: challenges and opportunities. Proc. IEEE **108**(2), 308–323 (2020)
22. X. Shen, X. Cheng, L. Yang, R. Zhang, B. Jiao, Data dissemination in VANETs: a scheduling approach. IEEE Trans. Intell. Transp. Syst. **15**(5), 2213–2223 (2014)
23. H. Peng, D. Li, K. Abboud, H. Zhou, H. Zhao, W. Zhuang, X. Shen, Performance analysis of IEEE 802.11p DCF for multiplatooning communications with autonomous vehicles. IEEE Trans. Vehic. Technol. **66**(3), 2485–2498 (2017)
24. H.A. Omar, W. Zhuang, L. Li, VeMAC: a TDMA-based MAC protocol for reliable broadcast in VANETs. IEEE Trans. Mobile Comput. **12**(9), 1724–1736 (2013)
25. X. Jiang, D.H.C. Du, PTMAC: a prediction-based TDMA MAC protocol for reducing packet collisions in VANET. IEEE Trans. Veh. Technol. **65**(11), 9209–9223 (2016)
26. F. Lyu, H. Zhu, H. Zhou, L. Qian, W. Xu, M. Li, X. Shen, MoMAC: mobility-aware and collision-avoidance MAC for safety applications in VANETs. IEEE Trans. Veh. Technol. **67**(11), 10590–10602 (2018)
27. DSRC Committee, Dedicated short range communications (DSRC) message set dictionary. Society of Automotive Engineers, Warrendale, PA, Technical Report J2735_200911 (2009)
28. NHTSA, 2016 Summary of Motor Vehicle Crashes. https:crashstats.nhtsa.dot.gov/Api/Public/ViewPublication/812580. Access Sept 2018
29. DLR Institute of Transportation Systems, Sumo: Simulation of Urban mobility. https:www.dlr.de/ts/en/desktopdefault.aspx/tabid-1213/. Access Sept 2017
30. CAMP Vehicle Safety Communications Consortium and others, Vehicle safety communications project: Task 3 final report: Identify intelligent vehicle safety applications enabled by DSRC, in *National Highway Traffic Safety Administration, US Department of Transportation, Washington DC* (2005)
31. F. Lyu, H. Zhu, H. Zhou, W. Xu, N. Zhang, M. Li, X. Shen, SS-MAC: a novel time slot-sharing MAC for safety messages broadcasting in VANETs. IEEE Trans. Veh. Technol. **67**(4), 3586–3597 (2018)
32. J. He, L. Cai, J. Pan, P. Cheng, Delay analysis and routing for two-dimensional VANETs using carry-and-forward mechanism. IEEE Trans. Mobile Comput. **16**(7), 1830–1841 (2017)
33. H. Kellerer, U. Pferschy, D. Pisinger, The multiple-choice Knapsack problem, in *Knapsack Problems* (Springer, Berlin, 2004), pp. 317–347

Chapter 8
Summary and Future Directions

In this chapter, we summarize the main contributions of this monograph and provide some future potential research directions.

8.1 Summary

In this monograph, we have introduced the vehicular network in supporting road-safety applications, highlighted networking challenges, and proposed the corresponding solutions at the MAC, link, and network layers, to guarantee low-latency and reliable broadcast communications for road-safety applications. Specifically, we have proposed and designed the following protocols/schemes.

- We have proposed a mobility-aware TDMA-based MAC protocol, named *MoMAC*, to reduce transmission collisions among moving vehicles. We have first identified two common mobility scenarios that can result in massive transmission collisions in vehicular environments. A simple yet effective slot assignment scheme then has been proposed, which can fully utilize the underlying road topology and lane layout information, to reply to the potential communication demands per vehicular mobilities. To eliminate the hidden terminal problem, *MoMAC* adopts a fully distributed slot access and collision detection scheme. Theoretical analysis and extensive simulation results have been carried out to demonstrate the efficiency of *MoMAC*.
- We have proposed a novel time slot-sharing MAC, named *SS-MAC*, to support diverse beaconing rates for road-safety applications. In specific, we have first introduced a circular recording queue to perceive occupancy states of time slots online, and then devised a distributed time slot sharing approach called DTSS to share a specific time slot efficiently. In addition, we have developed the random index first fit algorithm, named RIFF, to assist vehicles in selecting a suitable time slot for sharing with maximizing the resource utilization of the network. We have

© Springer Nature Switzerland AG 2020

F. Lyu et al., *Vehicular Networking for Road Safety*, Wireless Networks,

https://doi.org/10.1007/978-3-030-51229-3_8

theoretically proved the efficacy of DTSS algorithm and evaluated the efficiency of RIFF algorithm by using Matlab simulations. Finally, under various driving scenarios and resource conditions, we have conducted extensive implementation simulations to demonstrate the efficiency of *SS-MAC*. Note that, the proposed *MoMAC* and *SS-MAC* can work collectively to provide collision-free/reliable, scalable, and efficient medium access for moving and distributed vehicles.

- We have studied the 802.11p-based V2V communications in urban environments based on real-world data traces. We have the following two major insights. First, 802.11p works very reliably in urban settings with a wide range of "perfect zone" found. Second, LoS and NLoS channel conditions are crucial for reliable V2V communications. In particular, they have very opposite characteristics with respect to the PIR and PIL time distributions. The intervals between a pair of successfully received packets have an exponential distribution in LoS conditions but turns out to be a power law when in NLoS conditions. In addition, we have organized a discussion on how to utilize the unique characteristics of V2V link behaviors in order to improve the performance of vehicular network applications.

- We have investigated the link-aware reliable broadcasting scheme design for road-safety applications. Particularly, we have proposed *CoBe* to enhance the broadcast reliability by coping with harsh NLoS conditions, which integrates three major components: (1) online NLoS detection; (2) link status exchange; and (3) beaconing with helpers. Additionally, we have built a two-state Markov chain to model the link communication behaviors in regard to LoS/NLoS conditions, based on which we have carried out the theoretical performance analysis of *CoBe*. Extensive trace-driven simulations have been conducted, and both simulation and theoretical numerical results have demonstrated the efficacy of *CoBe* in terms of broadcasting reliability. In addition to the link-aware broadcasting scheme, we also have presented a case study of efficient unicast scheme for non-road-safety applications, which can enhance the spectrum utilization by avoiding inadequate opportunities for data transmission in advance.

- We have proposed a distributed adaptive beacon control scheme, named *ABC*, to conduct safety-aware beacon rate adaptation for vehicles under highly-dynamic vehicular environments. Three novel techniques have been integrated in *ABC*: (1) online congestion detection; (2) distributed beacon rate adaptation; and (3) adaptation results informing. Performance analysis on the efficiency of the devised algorithm has been provided. Also, we have implemented our proposed *ABC* scheme under SUMO-generated traces and conducted extensive simulations to demonstrate its efficacy. By adopting *ABC*, the beaconing reliability can be guaranteed even when the vehicle density becomes quite heavy, and meanwhile vehicles can adapt sufficient beaconing rates in accordance with the road safety demand in order to avoid the rear-end collisions.

8.2   Future Research Directions

In this monograph, we focus on the DSRC technologies to support road-safety applications, mainly investigating the delay and reliability performance of small-size packet delivery. Our future work is to integrate other spectrum resources, e.g., cellular networks, to extend our networking technologies to support multifarious vehicular services, such as location-based services, cloud-based comfort infotainment, and autonomous driving. We close this monograph with a discussion of some future potential research directions in this field.

- **Space-Air-Ground Integrated Vehicular Network (SAGVN).** To ably enable future automotive services, modern connected vehicles require an extremely versatile network that can simultaneously guarantee ultra-reliability and low-latency communications for assisted/autonomous driving, deliver high-bandwidth in-car entertainment, and support super-rate for VR/AR, etc. It is formidable to achieve satisfying service quality solely based on the terrestrial network due to the coverage holes and rigid resource feeding mechanism. Recently, the SAGVN has been proposed, which is widely regarded as a promising beyond 5G (B5G) vehicular network architecture. In SAGVN, in addition to the terrestrial network, the satellite network can provide ubiquitous wireless coverage, and the on-demand deployment of aerial network (e.g., unmanned aerial vehicles (UAVs)) can make the network adaptable to diverse environments. However, the integrated and heterogeneous network architecture can render the operation and management of network resources intricate, since different domain resources have distinct access technologies and variation features, which needs systematic and in-depth researches to well support vehicular applications.
- **Software-Defined Networking (SDN).** Considering the ever-increasing data demand and the service cost of networks, there will be various network resources, including DSRC, WiFi, TVWS, UAV, and satellite in the network, to support multifarious applications. The increasing number of vehicles and other communicating devices, high vehicle mobility, and rigid service requirements make the network management intractable. SDN-enabled technologies can offer potential solutions to achieve flexible and automated network management, efficient network resource orchestration, and global network optimization with cost-effectiveness, which has been envisioned as a key enabler to future vehicular networks. However, many issues such as the control architecture, online management algorithm, and slicing scheme, remain unclear.
- **Mobile Edge Computing (MEC).** MEC is an emerging architecture that can enable Internet and cloud-computing capabilities at the edge of the network, which can reduce network congestion and enhance application performance with executing computing tasks closer to users. This paradigm can facilitate extensive user-friendly and augmented reality applications, such as autonomous driving and VR/AR applications, where heavy-computing tasks can be processed within

a stringent time requirement. However, in vehicular networks, it is challenging to achieve satisfying performance of MEC, as service migration, intermittent connectivity, and bursty computing tasks can frequently happen when vehicle moving fast. Constrained by those practical challenges, the research on vehicular MEC is still in its infant stage, which needs further investigation.

Printed in the United States
by Baker & Taylor Publisher Services